Our Nature

QH
81
.G483
1986

Bil Gilbert

Our Nature

University of Nebraska Press

Lincoln and London

Copyright 1986 by the
University of Nebraska Press
All rights reserved
Manufactured in the United
States of America
Acknowledgments for the
use of copyrighted material
appear on page viii.

The paper in this book meets
the minimum requirements of
American National Standard
for Information Sciences –
Permanence of Paper for
Printed Library Materials,
ANSI Z39.48-1984.

Library of Congress Cataloging
in Publication Data

Gilbert, Bil.
Our nature.

1. Natural history –
Addresses, essays, lectures.
I. Title.
QH81.G483   1986
508   85-20891
ISBN 0-8032-2123-1
alkaline paper

For the three weird
but basically wonderful sisters—
Lee, Kate, & Lyn Gilbert.
May you appreciate and cherish the
above order of introduction.

Contents

Acknowledgments, viii

The Quick and the Dead, 1

After Franklin, 3

The Chicago Parrots, 31

The Curiosity of Thomas Nuttall, 47

Trophy Foraging, 75

Conceits and Feats, 91

The Great American Creeple, 103

Ghost Waters, 115

Cold, 135

The Original Dude, 153

Messing-Around Rivers, 169

Reflections on Hawking, 185

"God creates, Linnaeus arranges," 209

Great Bears and People, 229

Nature Loving, 249

ACKNOWLEDGMENTS

The following articles are reprinted courtesy of *Sports Illustrated* from the issues indicated:

"After Franklin," originally published as "Haunting the Arctic," July 8, 1974. © 1974 Time Inc.

"The Chicago Parrots," originally published as "Look What's Holed Up in Chicago," January 9, 1984. © 1984 Time Inc.

"Trophy Foraging," originally published as "Nibbling Bark Isn't Foraging," March 27, 1978. © 1978 Time Inc.

"The Great American Creeple," originally published as "Once upon a Time," October 21, 1974. © 1974 Time Inc.

"Cold," originally published as "Facing Old King Cold," March 14, 1977. © 1977 Time Inc.

"The Original Dude," originally published as "Thar Was the Old Grit in Him," January 17, 1983. © 1983 Time Inc.

"Messing-Around Rivers," originally published as "Streams of Contentiousness," June 28, 1982. © 1982 Time Inc.

"Great Bears and People," originally published as "Can We Live in Peace with the Grizzly?" July 23, 1984. © 1984 Time Inc.

The following articles are reprinted courtesy of *Audubon* from the issues indicated:

"The Curiosity of Thomas Nuttall," originally published as "A Somewhat Peculiar Fellow," September 1979. Copyright 1979 by Bil Gilbert.

"Reflections on Hawking," May 1985. Copyright 1985 by Bil Gilbert.

"God creates, Linnaeus arranges," originally published as "The Obscure Fame of Carl Linnaeus," September 1984. Copyright 1984 by Bil Gilbert.

"Nature Loving," originally published as "Reflections on Nature Loving," November 1976. Copyright 1976 by Bil Gilbert.

A portion of "Conceits and Feats" first appeared as "Big Hawk Chief, a Pawnee Runner" in *American West* magazine, July/August 1984, and is used with permission.

The Quick and the Dead

It is widely believed that we are to a large degree the products of our social history—the feelings, thoughts and actions of other people, known and unknown to us, both quick and dead ones—and of our natural history—inhuman phenomena that have been and are. I have come to accept this as a reality and to find it the most interesting of worldly ones. In consequence, as a writer I have been gnawing away publicly at the subject for thirty-five years.

The autobiographical content of the following pieces, remarks about things that have especially moved and made me, is incidental. It is a device for bringing up matters that are intended to be of more common interest: the extent to which nature, ours and the rest of it, is the product of running water, parrots, bears, James Reuel Smith, Carl Linnaeus, cold, and an infinity of other things.

Take Linnaeus as an example of how I think this works. In my own case the influence of the peculiar eighteenth-century Swedish savant is fairly easy to identify. My father was a botanist. He and thus I would have been different men had Linnaeus not been what he was, when he was. But as I understand it there is much more. Though not many now remember his works or even name, Linnaeus helped to shape the way we name and therefore think about all worldly things and thus since 1750 has influenced virtually everyone—certainly in the Western world and a good many other places as well. By way of illustration, it seems to me that what might be called the Linnaean element is a substantive one in how we now feel about national parks and deal with grizzly bears. Therefore the nature of certain Ozark rivers and Montana bears, and everything in between, is somewhat connected to and has been altered by the nature of Carl Linnaeus. That I can and want to isolate the Linnaean influence in myself may be a bit idiosyncratic, but I think the influence itself is general.

The conviction that we are all made of some of the same stuff and are in this thing together has led to the following ecological conceit. I sometimes have the notion that we quick things, animal, vegetable, and mineral, are members of an unimaginably large and complex relay team. We, the now existent, are carrying a baton that we have taken from things past and will soon pass on to future ones. I have never been able and (therefore) inclined to brood about the nature of the baton, where we are running to, or if there is an end to or rewards for the endeavor. However, the idea of being engaged in such a grand, mutual activity has been stimulating and soothing.

This line of observation and reasoning can also promote a certain sense of individual responsibility, that is, give reason for trying to behave one way rather than another. If we are the products of infinitely numerous, various, and ancient influences, we are also the producers of influences that will ripple along for no one knows how far into the future. Our times will very shortly be historic ones, our works and possessions antiques, and we will be ancestors. We will never know what kind of influences we will be, but trying to be, according to our lights, better rather than worse ones would seem to be a meaningful response for beings in our position. At the very least it would be a great embarrassment to be the ones who flat out dropped the baton. This might also be an absolute disaster, but that gets into cosmic considerations, which, as I have noted, are not for me.

After Franklin

John Franklin was born in 1786 in Devonshire. Thereafter he fought at Copenhagen, Trafalgar, New Orleans; was commissioned a captain in the Royal Navy, appointed governor of Tasmania, was knighted by William IV. He died in the polar ice pack, in 1847, on or near King William Island. In between these times and events he led three major expeditions of discovery into the North American arctic. It is because of those explorations that Franklin is remembered by some, that the name Franklin still appears frequently on maps of the high arctic, affixed to heights of land, bodies of water, lonely outposts.

Between 1819 and 1825 Franklin commanded two separate parties which between them traversed a thousand hard, virgin, overland miles between Great Slave Lake and the Arctic Ocean, and on the ocean paddled canoes, sailed small boats, laid down the first maps along fourteen hundred miles of coastline between Cape Beechey in what is now Alaska and Bathurst Inlet, to the north of Hudson Bay. In the 1840s, commanding a British arctic flotilla, Franklin was principally responsible (by his own efforts and because of the efforts of others who set out to rescue him) for the discovery of the Northwest Passage, the possible existence of which had intrigued bold men, driven them north for three centuries.

In comparison with other large figures of North American exploration—Coronado, Hudson, LaSalle, Lewis and Clark, Mackenzie—Franklin was personally rather commonplace. He was not a man of exceptional strength or stamina, scientific training or intuitive wilderness wit. His men did not seem to follow him out of great love nor be driven by fear and hate of him. As a leader he was principally a nagger and there was a strong fussy, even priggish streak in his character. What John Franklin was beyond everything else was an archetypal nineteenth-century British gentleman-adventurer. His activities and accomplishments stand as a monument to the power, or at least the former power, of persistence, doggedness, the efficacy of muddling through as an action tactic.

Without special historical interests or objectives, no one now contemplating a trip between St. Louis and Portland, Oregon, would dig out the Lewis and Clark journals to facilitate planning. However, if one for some reason wishes to journey overland from Great Slave lake to the Arctic Ocean it makes a good deal of sense to turn to John Franklin's *Narrative of a Journey to the Shores of the Polar Sea in the Years 1819, 20, 21, and 22.* It is one of the few published volumes dealing with that country. Also, because the central arctic, the demands it makes on men, has changed so little in the past 150 years, the *Narrative* still contains much pertinent information.

It was for this reason that six of us, having decided to seek stimulation in the north, came to John Franklin. (The six were Dick, John, Ky, Sam, and Terry, in their early twenties, and Bil, twice twenty and then some.) In the beginning we went to the *Narrative* for information on water conditions, portages, weather, the availability of accommodations, the possibilities of foraging for food. From there the decision to follow Franklin's route as closely and for as long as possible was an obvious one. When the planning and reading were finished and we were deep in the same fast water, muskeg, fly clouds Franklin had been in, our association with him and his men took a surprisingly intimate turn. We began to understand the Franklin party, their triumphs and frustrations, aches, pains and pleasure, not just in our heads but in our backs, hides, stomachs. Keeping up with them became a compulsive, competitive objective. In certain peculiar moments it seemed almost as if Franklin and his people were just ahead of us in time and space, that if we could sustain an extraordinary effort we might catch them at the next portage. If we did, we would be able to make some good medicine with these ghosts concerning matters that no contemporaries of either party could properly understand.

On his journey to the Polar Sea, Franklin left England on May 23, 1819, and did not return for forty-two months. The expedition was sponsored by the Admiralty and the Royal Geographic Society. His orders were to get himself to Great Slave Lake, which then was on the rim of the known world. From there he was to proceed north, striking, it was hoped, the Arctic Ocean at the mouth of the Coppermine River, which had been reached thirty-five years before by one Samuel Hearne traveling westward from Hudson Bay with a party of Cree Indians. He was then to explore eastward along the shoreline. Franklin was to record the latitude and longitude of "every remarkable spot," take temperature readings three times a day, make general meteorological, anthropological, and natural history observations.

Franklin was assigned three officers: a "surgeon" (chief naturalist), Dr. John Richardson, and two midshipmen, George Back and Robert Hood. Franklin was to employ voyageurs, guides, interpreters, other men when needed and where he could find them. In the main he was to depend upon the Hudson's Bay and North West companies, the two great fur corporations that maintained outposts across northern Canada, for the bulk of his provisions and equipment.

Franklin and his party arrived on September 9 at York Factory, then the principal fur depot at the southern end of Hudson Bay. From there traveling as everyone traveled in that time and place, by canoe, they set off into the interior, following the trade route up the Hayes River to the first Hudson's Bay Company post, Oxford House; then on to Norway House on Lake Winnipeg. In mid-October they came to the Cumberland Lake post on the lower Saskatchewan River, where they were caught by the freeze and overwintered. In June when the ice broke they continued on north by way of Isle de Crosse, Lake Athabaska, and the Slave River. In mid-July they reached Fort Providence on Yellowknife Bay on the north shore of Great Slave Lake, some seventeen hundred miles from York Factory. It was a true wilderness journey, but since it followed the existing canoe-trade roads, it was considered something of a milk run. That it took Franklin parts of two summers, ninety-one days, to make the trip was an indication that he was a green commander, inexperienced in the northern bush. (A veteran crew of voyageurs could drive a thirty-foot freighter canoe carrying two tons of cargo three thousand miles in a three-month summer season. It was a system of freight hauling that was not equaled for speed or efficiency until the transcontinental railroad.)

Even if the canoe roads of the fur empire still existed, even had we been so inclined, other considerations did not permit us to paddle from Hudson Bay to Great Slave. With two rather than forty-two months available, the best it seemed we could do was pick up Franklin's 154-year-old trail at the mouth of the Yellowknife. We traveled to this rendezvous as it is now customary to travel in North America during the summer months—by highway and van. There were certain hardships and annoyances, venal mechanics, exhaust fumes, and greasy roadside food, but this trip too was a milk run.

When Franklin arrived on July 28, 1820, Fort Providence, a North West Company establishment, was the most northerly outpost of the fur trade, sitting on the edge of some of the most inhospitable land, so far as humans are concerned, in the world. North of Great Slave the country is principally covered by glacier-tortured slabs, dikes, ridges of rock, some of which rise fifteen hundred feet above sea level, and by cold, deep waters trapped in basins of rock also gouged out by glaciers. Between the water and the rock there is little proper soil, only, as a rule, pockets of glacial sand and muskeg which floats on top of the permafrost. Surface water is very plentiful because drainage is poor and evaporation minimal. However, the land receives little rain, only about the same, ten or twelve inches a year, as the Arizona desert. It is one of the coldest places in the world, the mean (in many senses) January temperature on Yellowknife Bay being -17° and -70° readings can be expected now and then.

In places black spruce, white birch, alder, and willow grow grimly, hanging onto the rocks, but where it is found the forest is scraggly and grows very slowly. After a few hundred miles the trees give up and the country is then very aptly called The Barren Lands. With the exception of migratory species such as waterfowl and caribou who escape in the worst parts of the year, warm-blooded creatures, like the plants, are thinly distributed, hang on precariously.

Fort Providence was built in the 1790s with the notion of establishing a trade with the Dog Rib Indians to the northwest of the big lake and the Copper Indians to the northeast. Commercially it was a bad idea, because these peoples, of whom there then may have been three thousand or so, were among the poorest on earth. The coming of the traders did little to uplift them. Furbearers were not plentiful in their country and the Indians were not skilled or energetic trappers. Occasionally they would accumulate a surplus of caribou or fish which would be shipped to more southerly posts, but jerked caribou and smoked fish were not high-profit items.

In some respects social conditions around Yellowknife Bay have changed considerably since 1820. A few years after Franklin passed by, Fort Providence was mercifully abandoned. In time gold was discovered in the area and the mining camp evolved into the city of Yellowknife. Yellowknife is now the metropolis of the central arctic, being the capital of Canada's Northwest Territories, a vast tract half the size of the United States but inhabited by only thirty-five thousand people, a quarter of whom live in the capital. Mining is still the principal industry, but increasingly Yellowknife is becoming a service community, the headquarters of various federal social agencies, the air supply base and communications center for isolated public and private posts scattered across the ice and tundra virtually all the way to the North Pole. Like any busy, civilized place, Yellowknife now has its traffic problems, dope dealers, Kentucky Fried Chicken.

Despite the differences there is a fundamental similarity between old Fort Providence and the new Yellowknife. The former was, the latter is, the end of all roads. In Franklin's day, the established canoe trails that extended from the Atlantic to the Pacific went no farther north than Fort Providence. Now the even more intricate web of roads that connects the Garden State Parkway to the Santa Monica Freeway and makes everything in between a segment of the same road begins to dwindle, funnels into one road in northern Alberta. This is a shallow gravel pit, loosely organized into something called the Mackenzie Highway. It crosses the Mackenzie River (by ferry in the summer, on ice in the winter), skirts the northwestern shore of Great Slave, and reaches Yellowknife seven hundred hard miles from the nearest pavement. It seems, considering the significance, that where it stops there should be a memorial, a fancy arch, or at least a bronze plaque suitably inscribed "Here is where all roads end." In fact, the last of all roads peters out in a cluster of picnic tables and rustic toilets, at a territorial campground a few miles outside Yellowknife.

Upon reaching such a dead end it is reasonable in any year to make local inquiries. Arriving at Fort Providence, Franklin got in touch with a Copper Indian chief, Akaitcho, who led a band of some fifty souls. It was one of the shrewdest moves he made in all of the forty-two months, for Akaitcho was by any standards a good man and, considering the place from which he came, an extraordinary man, "evincing," as Franklin wrote, "much penetration and intelligence."

Akaitcho, Franklin, and their people met at Fort Providence on July 30 and made medicine for a day. Drastically edited, the negotiations went more or less as follows. Akaitcho said that he had heard rumors that Franklin was a great medicine man who could restore the dead to life, and if so, he was of course honored to meet someone of such rare and useful talents. Franklin replied that the rumors were untrue, that he was an explorer anxious to find his way to the mouth of the Coppermine River. Akaitcho expressed regret that Franklin was not a death-defying shaman and inquired as to the nature of the exploring business. Franklin delivered an involved lecture on the power and glory of curiosity, saying he had come "to make discoveries for their [the Indians'] benefit as well as that of every other people." Amazingly, since he came from a culture that had so little experience with abstract frivolity, Akaitcho picked up on this idea, said that he could see how exploring might be amusing. Franklin then said that since he did not know how to get to the Coppermine, he would be delighted if Akaitcho and some of his people would guide him. Unfortunately, his—Franklin's—party was critically short of provisions, and if the Indians came along they would have to feed themselves and, he hoped, help feed the whites. By way of compensation, Franklin would give out some red cloth, a few pocketknives, ten obsolete muskets, and a keg of weak rum. Also he would see to it that Akaitcho's debts at Fort Providence were paid. Akaitcho mulled over the proposition— very likely considering the fact that it was as hard to live one place as another in his godforsaken country—and finally said yes, that he and all his people "would attend Franklin to the end of his journey and do their utmost to provide them with means of subsistence."

Having concluded their business arrangements, Akaitcho and Franklin cohosted a fete on the last day and night of July. Just as the rum began to flow nicely, someone noticed that Franklin's tent, in which he had left an untended smudge pot to ward off mosquitoes, was going up in flames. Franklin was mightily disturbed by the incident, having decided that to impress the Indians he must at all times deport himself with gravity and dignity, a hard pose to maintain when you have just carelessly burned down your quarters. However, Akaitcho took the matter in stride, telling Franklin, as he was to have occasion to tell him so often in succeeding months, to cheer up, it was the kind of thing that could happen to anyone. In fact the accident may have improved Franklin's position. The tent burning, which was regularly followed by similar comic to tragic blunders, seems to have convinced the remarkable Akaitcho that he had on his hands a well-intentioned but incompetent bungler and that it was his compassionate responsibility to do what he could to protect this Fool of God.

The Indian population is now larger in the vicinity of Yellowknife than it was in 1820. However, these are no longer bush people as they were in the days of Akaitcho. The establishment of the fur posts offered the original residents an alternative to their traditional subsistence style of living. They chose it gladly; commenced exchanging furs for the marvelously useful and interesting goods that the traders offered. In time they became largely dependent on imported supplies. In the twentieth century the fur business declined and with very few exceptions the Indians came in to Yellowknife and other European-style communities, where they have generally been supported by bad work and various forms of public assistance. Meager as the rewards have been, they apparently—for many practical and psychological reasons—seemed greater than those their ancestors enjoyed in the Arctic wilderness. (Alcohol is a very frequent reward/escape.) The subsistence skills and attitudes have therefore largely disappeared.

White bush expertise, never extensive, has also declined in recent times, particularly since the advent of the airplane, which makes it possible to drop down out of the sky after breakfast, do a little fishing or prospecting, and be back in Yellowknife for dinner. In consequence what intelligence was available tended to be generalized, second hand. The most practical advice was offered by a lady newspaper reporter who said that Adolph's Meat Tenderizer is a good thing for easing the discomfort of mosquito bites. It is. Another lady, with whom a junior member of the expedition became acquainted, passed on the information that should we reach the village at the mouth of the Coppermine we would find a man there in possession of two pounds of good hash. A sergeant of the Royal Mounted Police made notes on our proposed route, color of the tents, addresses of next of kin, in case search and rescue operations should become necessary. A territorial public affairs man said that on the day we embarked he would drive the van back to Yellowknife and store it in a government garage. This was not as grand a gesture as Akaitcho made to Franklin, but it was a generous and useful one.

Having made such medicine as we could, we did as Franklin had done, had a wing-ding on the last night before leaving the road. Again the natives proved to be understanding.

"Are all your chaps in town?"

"All of them, Sergeant, ever las one."

"Is one of them a large chap with girlish hair?"

"Very lurg, very gurlish?"

"Yes."

"That is Sam. What has he done?"

"No trouble. Some of the men have seen him. He looks like the kind of chap who might make trouble."

"Sergeant, we are engaged in revling. We revelly preciate your corcern and contesy. I mean we preely reciate your cortest and cern."

"You chaps leaving in the morning?"

"First train north."

"We certainly hope you make it and have a grand trip."

After consulting with Akaitcho and his men (none of whom had personally traversed all of the proposed route) Franklin decided to start north by going up the Yellowknife River, which the Indians called the Begholodessy. The Yellowknife, like many of the rivers that flow out of the central arctic, is not a conventional stream. It is, in fact, a 150-mile-long string of bogs, ponds, small and large lakes connected by relatively short inlet and outlet channels. Therefore, going upstream is not a matter of continually fighting the current, since where it widens to lake size there is no discernible movement of the water. However, to compensate there is a great deal of current and worse in the connecting links. Where a ten-by-three-mile, 200-foot-deep lake, of which there are many, pours out of its rock basin down into the lake below, there is inevitable turbulence, cataracts and falls. By tacking across the current, making use of back eddies, it is sometimes possible to paddle partway up these channels, but there always comes a point where it is necessary to get out and either wade, dragging canoes by ropes through the rapids, or land and portage the boats and gear around the impassable water.

From Great Slave going northward there is a short stretch of more or less true river, a moderate series of rapids and the first of the big interior lakes, Lake Prosperous, so named (as were most of the other landmarks along the route) by Franklin. Prosperous is seven miles long, a mile or so wide, studded with rocky islands, bordered by sharp cliffs and crags that are lightly covered with spruce and birch. Franklin got to Lake Prosperous on August 2 with his English officers, twenty-four Canadian voyageurs and guides, and two women the voyageurs brought along to cook, mend, and what have you. They traveled in three twenty-eight-foot canoes and were accompanied by seventeen smaller Indian canoes in which floated Akaitcho and all of his people.

It was a pretty day, clear, balmy, temperature in the mid-sixties, with fleecy white clouds floating overhead, according to a pastoral sketch made by Robert Hood, who shared with George Back the job of expedition artist. (One of the pleasant surprises of the central arctic is the summer weather. Being dry, it is invariably sunny, and at these latitudes the sun shines for twenty hours or so a day. The twilights can be nippy, but in general the temperature is just right for heavy physical labor.)

Of the first day out, Franklin wrote in the *Narrative,* "We are all in high spirits being heartily glad that the time had at length arrived when our course was to be directed toward the Coppermine River and through a line of country which had not been previously visited by any Europeans."

There comes a time in any such enterprise when for better or worse the die is indisputably cast, when all doubts and reservations no longer signify, when it is no longer necessary to explain motives and plans to well or bad wishers. Like Franklin, we were heartily glad to have come to the casting-off point.

The first major portage on the Yellowknife comes at the end of Lake Prosperous. It involved a hundred-foot climb and a three-quarters-mile carry around a series of cascades that flow out of Bluefish Lake to the north. The logistics of any portage are simple, and so is the work, simple horse work. In our case we had in the beginning twelve hundred pounds of gear (eating and use reduced this total as the summer progressed). This meant that each of the six had to make two round trips at each portage, i.e., there were nine 130–140-pound pack loads, three seventeen-foot aluminum canoes to be carried. All of which can be, had frequently been, calculated in the abstract. However, no matter how extensive and cunning the pre-planning, there comes a time when truth—what sacks will mate nicely on the pack board, which ones will remain forever incompatible no matter how lashed or cursed—can only be found by thrashing around in the water, mud, and on the rocks.

The portage out of Prosperous Lake was an exhausting, temper-frazzling, memorable one, because by nature it is a hard carry, because it was our first, because we were not yet intimately acquainted with the hauling properties of our gear, and most particularly because it was here that we first went into full combat with the Curse of the North—the bugs. Southerners, thinking of picnic ants, patio mosquitoes, golf course gnats, may believe otherwise, but it is not possible to exaggerate insect power and insect agony north of the 50th parallel. In the short summer season, midges, blackflies, deerflies, mooseflies, mosquitoes hatch out by the billions. Outnumbering their potential victims and food sources as they do, each member of the predatory armada seems always to be in a kind of blood frenzy. They probe, puncture, wound, choke, poison the body, weaken the mind. Now and then when a man goes down alone, unprotected in the bush, they kill, suck out his life drop by drop.

John Franklin was if nothing else a pedantic man. Consider then his *Narrative* after a bad bug day. "The horse [moose] fly ranged in the hottest glare of the sun and carried off pieces of flesh at each attack . . . the smallest, but not least formidable was the sandfly. . . . The food of the musquito is blood, which it can extract by piercing the hide of a buffalo. The wound does not swell like that of the African musquito but it is infinitely more painful and when multiplied a hundred fold and continued for so many successive days it becomes an evil of such magnitude that cold, famine, and every other concomitant of an inhospitable climate, must yield the pre-eminence to it. . . . The musquitoes swarmed under our blankets, goring us with their envenomed trunks, and steeping our clothes in blood. We rose at daylight in a fever, and our misery was unmitigated during our whole stay.''

So far as northern bush travel is concerned it seemed to us that the most important technical advance made between Franklin's day and ours was the development of good mosquito bar and better, if by no means perfect, insect repellents. It was the ability of Franklin's people to keep going, vulnerable as they were to the bugs, which impressed us more than any other of their physical feats or accomplishments.

The Lake Prosperous portage was bugwise an average one, but shocking, again because it was the first. From Prosperous the carry is up over a rocky ridge, then down for a half mile to Bluefish through heavy thickets of birch and patches of muskeg. Between the ridgetop and Bluefish there is a gauntlet of ferocious insects. When we passed by, grand armies of mosquitoes and deerflies, stiffened by a few heavy divisions of mooseflies, were on duty. Purchased, packed in Pennsylvania, a beekeeper's headnet seems to be a curious, even comical item. In the bush between Prosperous and Bluefish a man will fling off a 130-pound pack that he has laboriously raised, tear open the lashings, paw wildly through his possessions in search of a headnet, which at that moment seems a more useful and desirable invention than smokeless powder, the combustion engine, or cola. Nothing yet devised totally defeats northern bugs, but a head net filters them somewhat, at least keeps them from matting hair and beards, clogging eyes, nostrils, ears, throats.

We camped the first night, as we would on many nights, where Franklin had camped before us. The choice was not made for the sake of historical high fidelity, but was more or less dictated by environmental imperatives. Anybody coming out of the Prosperous-Bluefish portage late in the day would be inclined to camp where Franklin did and we did, on a flat, rocky island, unless they were masochists or fools. The place has a nice loading and unloading spot, but the imperative thing about it is that it sits half a mile or so out in the lake and has hardly any vegetative cover. Therefore it has about as few bugs as any place that can be found in this country.

In general there are three kinds of location in this land where the bugs are tolerable if not absent. They are: flat islands in the middle of biggish lakes; bare, rocky points jutting out into the water; and sandy hills called eskers (they were laid down by ancient rivers flowing under the ice). The critical common element is that all three are dry, more or less clear of low brush, and are open to the wind, which serves to discourage large clouds of bugs from hanging around and sucking blood. Fortunately, such bug sanctuaries are not uncommon. Normally in the course of a day's travel half a dozen sites will be passed which are satisfactory so far as insect security is concerned and are superb on the score of camping conveniences and necessities. There will be a sloping shelf of rock at the water's edge, hand cargo handling and washing; higher up wide, flat ledges upholstered thickly with reindeer moss for sleeping. There is bone-dry driftwood in suitable lengths on the beaches and resinous birchbark, God's Own Firestarter, everywhere. Beyond and about everything else there is water, the true staff of life, whose absence or filth makes living in the open a chancy, exasperating, risky business elsewhere on the continent, even in so-called wilderness areas. In contrast, smelters, pulp mills, sewage plants, and other sources of high-tech garbage being so many miles away, the billions of gallons of water in the interior of the Northwest Territories lack the customary civilized contaminants. Also, because the deep, rocky basins support little plant life, because there is little soil and silt, rainfall and runoff, the water is of a clarity that astounds those accustomed to the murkier waters of the south. When we left the north at the end of our travels, the thing we immediately missed the most was being able to bend down anywhere and scoop up a drink of clear 40-degree water.

    The other great, abundant, useful resource of the country is also water-connected. The deep, cold lakes and rivers of the Northwest Territories constitute some of the world's most remarkable fish factories. At the first camp on the rocky island, Sam, who was the best of a very bad lot of sports, took out one of the two telescoping spinning rods we had brought along and on his second cast reeled in, to everyone's astonishment, a ten-pound pike. Shortly he repeated the performance and then turned the rod over to others who had suddenly become believers. Not only did we catch fish, but more important from the standpoint of skeptical non-anglers, we could watch ourselves catching them. Even on nonproductive casts, large torpedo-like shapes, often several at the same time, would rise out of the deep water, track the lure toward the island. In an hour we took enough pike for three meals for the six of us and perhaps two weeks' food for a mink family, a female and two kits who lived on the island and kept busy dragging fishheads and guts into their burrow.

As it turned out, this first one was not an unusual place or experience. In the large lakes and river channels, pike and lake trout, up to thirty-five pounds or so, were common. However, it takes some effort to haul them out of the water and it is hardish work to slaughter and dress these lunkers. From our standpoint, altogether the most admirable arctic fish was the grayling, often called the bluefish.

Grayling, who only inhabit the arctic drainage system, are very attractive creatures of two or three pounds. With a large, fanlike dorsal fin, an iridescent hide that flashes with blues, golds, greens, reds, they resemble large Siamese fighting fish. Not only do they look well, but as aghast sporting friends were later to tell us, grayling are also among the gamest of freshwater fish. We learned nothing about this facet of their character, since it was our custom to yank them out of the water on the same sort of line, with much the same non-technique, as we used for thirty-five-pound pike.

Grayling hang about in fast white water, feeding on larvae, nymphs, sculpons. Our practice was to fish for them at the head of rapids. They are also found at the foot of cascades, but there too are big pike and trout who wait, for among other things, grayling. There were usually plenty of the smaller fish to go around, but now and then while we were pulling in a grayling a pike would rise up and nip off half or more of our meal from the hook.

According to our values the best things about grayling were that they are very plentiful, can be cleaned in a matter of seconds, and are, we thought, tastier than either pike, trout, or whitefish. When we were particularly ravenous we would make a brief midday stop at a rapids and if all went well have within a half an hour a panful of grayling to go with our dried soup. All of which improved our strength and spirits.

By and large Franklin's party on a per capita basis ate less fish than we did. George Back, the midshipman, had packed a flyrod, but this was essentially a toy for amusing and amazing the Indians. The expedition also had gill nets, but they are the devil's own work to carry and set and as a rule were used only at long layovers. Most of the time the Franklin party had no fish or only a few fish scraps, which they scrambled for like starving alley cats.

In any enterprise that involves living and traveling where there are no settlements, no supply depots, no agriculture, the concern that dwarfs all others is eating. In general there are two ways to cope. The first is to bring food with you. The second is to forage. Neither method is entirely satisfactory, which is why the bulk of the world's population is found in settled districts that support agriculture and commerce. The disadvantage to carrying food is just that —it must be carried, picked up, put down, endlessly lashed and lugged. Foraged food does not have to be carried far (though foraging tools, guns, ammunition, nets, etc., do) but one way or another it must be hunted. Hunting is expensive in terms of energy (food) and is chancy.

The six of us opted mainly to carry. We brought with us food for six weeks, a bit over two pounds per man per day, rations which fortunately were bulked out by fish, occasionally blueberries, raspberries, strawberries. Except for some jugs of honey, peanut butter, and oil, the provisions were dry in the interests of saving weight and preservation: rice; noodles; wheat, soybean, corn flours; nuts; raisins; powdered juices; tea; and salt. Currently there is an exotic array of so-called freeze-dried foods that are touted as lightweight camp fare. They are good enough for affluent people to carry in motor homes, for impressing other out-of-car-doorsmen, but they are exasperating items to take beyond the end of the road. They are bulky though light, complicated to cook, hideously expensive, and no matter what may appear on the label—strawberry mousse, beef stroganoff, crab imperial—they all end up tasting like wallpaper paste with chemicals added.

In the beginning, Franklin apparently intended to do as we had done, that is, to carry most of his supplies. However, he found that commodities which could be casually listed in memos written in London simply did not exist or were in very short supply in the Canadian bush. Also, the fur factors tended to regard all newcomers from Europe, no matter what their credentials, as fools, fops, glory hunters, and were not overly eager to share supplies with them. Therefore when Franklin left Fort Providence he was badly underprovisioned. He had with him two casks of flour, two hundred caribou tongues, a little dried moose meat, a case of chocolate, two cannisters of tea, some dried soup and yarrow—all of which amounted to only ten days' worth of food for a party that was setting off for no one knew how many months in the wilderness. Franklin expected to overcome these obvious deficiencies by employing Akaitcho and his men as hunters. That was a theoretical but not a practical solution.

It is true that the Indians had lived, after a fashion, off this land for many years. However, they were few in number and survived by going to good food places (productive fishing holes or spots where they could intercept the migrating caribou herds), staying there, and spending nearly all their waking moments collecting food. Franklin expected the Indians not only to continue to feed themselves but find food for thirty extra mouths. At the same time, the party was to spend fourteen hours or so a day in the nonproductive (so far as food was concerned) labor of exploration. Akaitcho and his hunters gave it a valiant try but their poor land simply did not produce sufficient food surpluses to support such an enterprise in any sort of comfort. Again it was contrary to the environmental imperatives.

Since time immemorial there has been really only one course open to men who wanted to do something frivolous, such as find the mouth of the Coppermine River, in the central Arctic barren land that produce so little excess food; that is, of course, to go hungry. Franklin and his men learned of this technique very early. The food they brought from Fort Providence did not last ten days, only six. During the next twenty months there probably were not twenty days when they were not hungry. They now and then went foodless for two or three days. They often ate rock tripe, a large, ugly-looking lichen that can be boiled into an unappetizing, non-nutritional mucilaginous stew which fills the stomach to the gagging point. They ate carrion, boiled old bones, pack thongs, their boots and moccasins. Even so their travel-by-hunger technique was not completely successful. Four of the party ultimately starved to death, the deaths of others were hastened by malnutrition.

It would be obviously inaccurate and profoundly presumptuous to suggest that in our five weeks on their trail we suffered as Franklin's party did. However, the situation in regard to edibles has changed so little since 1820 that we had a kind of introduction to hunger, sufficient to enable us to sympathize seriously with the Franklin expedition. Our work day was twelve or fourteen hours long. We began it with a cup of dehydrated juice, a cup of tea, a cup of some sort of flour mush. In midday there was a cup of soup, one hardtack cracker, a half handful of nuts, raisins, sunflower seeds. At night we filled a two-gallon pot with rice, noodles, fish, whatever else we had come across, boiled the mess, and licked the pot clean.

Sam is a big, well over six feet, two-hundred pound, man and an energetic one blessedly given to taking the extra pack, making the extra portage trip. Also, in more hospitable places he is a notable feeder, a truly national-class junk food eater. "I am not exactly hungry," said Sam judiciously after one of the evening clean-the-pot feeds, "but I am not what I would call full. You know that stuff about not eating until you are uncomfortable. That is a bunch of New England–puritan crap. One of the most comfortable things in the world is being uncomfortable from eating."

On such a diet, doing such work for five weeks, Sam ceased to be a big man, became a tall, gaunt one, ending up thirty-five pounds lighter than he began. Bil, whose horizontal dimensions had as the years progressed begun to approach Sam's, left twenty-five pounds of himself somewhere on the upper reaches of the Yellowknife River. The others, having come with smaller surpluses, had less to lose, only ten or fifteen pounds apiece. On the whole Weight Watchers is an easier if less effective program.

We knew that we ate better than Franklin and his men had, thought that we were at least as strong as the men we were following. It seemed that we worked as hard and as long as they did. Nevertheless, we slowly but steadily lost ground to Franklin. We would work until ten at night, paddle four lakes, make five portages, and find at day's end when we looked at the *Narrative* that Franklin had taken his cumbersome, famished party two lakes and a cascade beyond us.

There were technical reasons for the greater speed of the earlier travelers. Our aluminum canoes handled better in fast water, were more durable than the birchbark ones Franklin used. However, on a long stretch of flat lake water, especially going into the wind, the big freighter canoes, paddled by a dozen men, went twice as fast. It was simply a matter of manpower.

More importantly and curiously, Franklin, even though he was the first to travel in these parts, went faster because he had a better idea than did we of where he was going each day. We were guided by the *Narrative* and a set of topographic maps. However, the *Narrative* is a general traveler's journal, not pilot's instructions, and the maps made by aerial survey are of a very small scale, 1:250,000, and therefore do not solve specific puzzles of terrain. Particularly where it was necessary to leave the river and head overland we often had to stop, thrash about in the brush to find and then clear portage trails. Franklin was spared much of this sort of work and delay because he had with him living topo maps in the persons of Akaitcho and his hunters. In many cases some of the Indians immediately knew where the trail led because they had hunted or trapped in the country. In new territories, the hunters, who often worked several days ahead of the main party, had found, picked out, and marked the best portage by the time Franklin came up with the heavy freight.

There was another, essentially metaphysical factor which cannot be translated into miles per day but which probably worked to our disadvantage. For example: toward the end of things Sam was had by microbes. His throat became badly infected, closed to the point that he could only sip liquids, a few drops at a time. He was shaky, feverish, beset by hallucinations; would find himself, for example, struggling to keep up with a weirdly mixed portage party of us, Franklin's men, and hands from the Pennsylvania farm where he worked. One morning about two o'clock, sitting watching the twilight and dawn mingle over the water, he whispered painfully.

"On top of the rest I had a revelation."

"Aha."

"I thought it was food I was craving—cheeseburgers, milkshakes, pizzas, all that. But I just figured it out. It isn't food. I want freedom."

"Aha."

"I mean we've been like parts of a machine. You can't do anything on your own because then everything breaks down. Everything we do has got to be done if we keep going. What I'm really craving is to be a free agent. The food was an excuse."

"It's the great freedom symbol. People bitch about the food in prison, in the army, in dorms. It isn't that it is so bad, it's because they have no choice about what or when or where. If you have to eat by the numbers you know you're trapped."

"Maybe; my revelation didn't go that far. But if I ever get back to Yellowknife, I'm going to buy all the food I can carry and get a room in that hotel and sit there and eat it all alone. I don't want any of you people in my movie for a couple of days."

"Good plan."

The problem was this: In 1820 the lives of Franklin's voyageurs and Indians, the men who did the work and enduring, were ordinarily short and brutal. The hardships of the polar expedition were only marginally more severe than those they experienced routinely. A few times a year they might get a square meal, have a chance for an orgy of drink or sex. Otherwise they were expected, themselves expected, to be hungry, cold, exhausted; to be bent under a pack, over a paddle, by another man's command and to function like a part of a mindless machine until they broke down like a machine part.

The difference was not so much that we have degenerated significantly. It was that during the intervening 150 years the level of comfort and human expectations has changed dramatically for those of our class. Too many soft sheets lay between Sam and Akaitcho. We were plagued by questions of why; by trying to find some point to, lesson in, the ordeal. Such speculation is counterproductive in these circumstances, gnawed away at our minds and bodies.

"And there is another thing that has changed. Franklin would say 'all right, chaps, you've done ten miles through the muskeg today. Carry on another five and you can have a nice mess of rock tripe and moccasins.' And they would do it because he said to and he was the man."

"He could say it because he sat on his ass all day while those poor bastards paddled and portaged and fed him."

"Maybe that's what a great expedition needs—a well-fed, rested fat cat who sits in the middle and keeps the chaps pressing on. He can be objective because he doesn't hurt.''

"What this expedition needs, old man, is for you to pick up that damn canoe and get your butt in gear.''

"And that time on Reindeer Lake when the chaps said they had to have some food or they wouldn't paddle. Franklin said he 'chided them' but he also said he would shoot them. Even way up here there is probably some law about shooting one of you freaks. Anyway there aren't enough of you.''

"What this younger generation needs is some discipline. Respect for authority.''

"Exactly.''

All of which more or less came to a head in a place called the Nine Lakes. A hundred miles or so north of Great Slave, the Yellowknife begins to narrow as it approaches its sources. It enters a twenty-mile-long gorge, up which, Akaitcho advised Franklin, it was impossible to paddle and dreadfully difficult even to drag canoes. The way, said Akaitcho, to get around the obstacle was to make a two-and-a-half-day portage through a string of nine small lakes that lay parallel to and west of the gorge. Accepting this advice, Franklin wrote of the first day in the Nine Lakes, "We crossed five portages, passing over a very bad road. The men were quite exhausted with fatigue and by 5 p.m. were obliged to camp.''

"We are in trouble,'' said Terry, who is a pragmatist and had begun to appreciate that Franklin was inclined to treat portage problems casually because he did so little work on them. "We are the men he is talking about,'' said Terry, getting to the heart of the matter.

Nearly all the components of a difficult portage are to be found in the Nine Lakes region. The road is indeed bad. The small lakes are separated by sharp, steep, rough ridges of rock. Between the ridges are swales of muskeg, full of hidden potholes that sink down to ice water and permafrost. Being comparatively sheltered, the intervale swamps support especially dense thickets of spruce, birch, and arctic rose. Finally, the term Nine Lakes is a figurative one. Franklin happened to use nine lakes but in fact there are some twenty bodies of water jammed together in a bewildering maze. We spent six hours on the first day cutting a trail through the arctic jungle to what we thought was the first portage lake. When we had made the carry to it, sat down to puzzle out the next move, we found that we had miscalculated our compass declension, had to cut another half mile of trail to get back to where we belonged. On the third day a howling wind and rainstorm came down out of the north, pinned us for a day on the shore of a lake, across which it would have been suicidal to attempt to paddle.

The eventual carry back to the river was a torturous one, beginning in the muskeg, dropping down over the wall of the gorge across ledges and loose scree. Coming back for their second trip, Sam and Bil sat to brood for a time before picking up their respective canoes.

"I have been thinking about something today."

"Good."

"I have been wondering if I am having any fun."

"Are you?"

"I don't think so. I think we are hung up with an idea like a segregationist or religious fanatic. We are killing ourselves trying to act out this movie — After Franklin."

"To me the fun is that we are seeing some country, doing some things that damn few people have or ever will see or do. The work is how we are paying. If you want to see Yellowstone or Tahiti you pay to go there, but so can anyone who has money. The only way you pay to get here is by half killing yourself. That is why we haven't seen anybody in a month. The prices are too high. Makes you feel rich."

"But that's what I mean. We're paying but we're not collecting. I've been thinking about this trip for two years, ever since we spread out those maps on your kitchen floor. But we really aren't seeing anything, getting to know anything. At night when we stop I think I should go off and look at the scenery, or find an animal or think about the meaning of it all, but I'm too goddamned tired and hungry to do anything but eat and sleep. We get up in the morning and do it again. Mostly what I see is the underside of the canoe when I'm carrying it through a swamp or down a cliff, that and the backside of Terry's head when we're paddling it. You probably think I am a baby, but that is how it is for all of us, even for you, and you are the real freak."

"That's not what I think. No offense intended; I think you have got a lot of guts and some brains."

"Thanks."

"Let's go on a little bit, a few days. We'll think about it, make some medicine when we get out of the gorge."

We carried out of the Yellowknife gorge into a big pair of interconnected lakes, the Carps. (The gray sucker was called carp by Canadians.) The weather had cleared, calmed, and warmed. On an afternoon we came up on a flat, parklike island set with small clumps of birch and spruce arranged as if they had been situated for purposes of ornamental design. There was a landing cove on the sheltered side of the island. Above the landing was an elevated table of rock that seemed to have been raised specifically as a tent platform. It was cushioned with moss, overlooked twenty miles of water and bush. We could not or would not pass this place.

(Coming out of the brutal Nine Lakes, Franklin also had stopped on the Carps for the same reason as we did and in much the same mood. "We determined on halting for a day or two to recruit our men of whom three were lame and several others had swelling legs.")

On Carp Island we slept late, splurged a little on food, sunbathed on the rock deck, puttered away at small, easy chores; washing clothes, mending boots, a leaky canoe, splicing ropes. We built an artful fireplace, fished for fun, went off in empty canoes to sightsee in the adjacent country, spent an afternoon lying on a sandy ridge watching a moose cow and calf feeding in a pond below, came upon a sow and cub grizzly in the berry bushes, a pair of courting martens.

In the course of things Terry, who in all such adventures is our main camp man, cataloged our remaining provisions. We had food left for two and a half weeks. (We had then been out three weeks. Franklin had covered the same 120 miles in thirteen days.) Seven days' travel to the north, at Franklin's pace, were the remains of Fort Enterprise, the collection of cabins which the Franklin party had built and in which they had spent the winter of 1820–21. From Fort Enterprise it had taken Franklin another thirty-four days to reach the polar sea. The distance was 334 miles, of which 117 were portage miles. Clearly we had neither the provisions nor time before the freeze to reach the ocean. We could by pushing on as we had been get to Fort Enterprise and by doing so duplicate Franklin's travels in the summer of 1820. However, on Carp Island, we did not think very seriously about going on to

Enterprise. Sam had spoken the truth. We had not come out to discover the Polar Sea, extend the limits of geographical knowledge or the realms of the king. Those things had been done at the proper time by the proper men. We had come for light reasons, sensual ones, to see, feel, enjoy an unusual place. Somehow we had lost our way, been caught up in an obsession, been infected with hubris that was steadily drying up our sense and senses. In trying to travel with ghosts we had fallen into bad company.

The three easy days on Carp Island were like waking on a sunny morning after a hard, frightening night of dreaming. There we parted with John Franklin. He and his lame and starving men drove on to the north. We turned and drifted south.

"We've got plenty of food now. We know the portages. We're not going to rush back so you people can eat cheeseburgers in Yellowknife. We'll mosey along, stop when we want, look around."

"That's good, but we are not going back through the Nine Lakes. We'll walk out or live here, but there is no way we are going back in there. What about running the gorge?"

"We don't know anything about it. Maybe we can run it, but maybe if it narrows down we'll come to a place where we can't go ahead because of the water and where the walls are too steep to climb. We'd have to drag the canoes back upstream and then still do the Nine Lakes."

"Let's take the chance."

"Why not?"

It could have been that way, a very bad road, but it was not. In fact—and this may have been the original discovery of the expedition—the Yellowknife gorge is an extraordinary place to paddle a canoe, an extraordinary place to be no matter what means is used to reach it. From the falls where the Yellowknife empties out of Lower Carp Lake south to the point where the Nine Lakes portage leaves the river is about twenty-five miles. The water is fast most of the distance, and a good paddler in a covered canoe without much gear could go through the gorge in a day. We took three days and thought afterward that it would have been better to take a week.

We paddled for a time between five-hundred-foot cliffs of dawn pink rocks that were mottled with big rosettes of sea green and black lichens. There was a broken cover of clouds overhead, occasional rain squalls, but between the clouds and the rain the sun was bright, made the chop of the current glitter gold and silver for as far ahead as we could see. High on the east cliff an adult bald eagle sat taking the sun, watching us as she perched on the rim of her bulky nest. From west to east a moose swam across the sparkling water ahead of us. For good measure a perfect double rainbow arched across the gorge.

"I'll be a son-of-a-bitch," said Sam. "Walt Disney has been here."

"My heart soars like an eagle," said John, who is one of the last surviving men who can say such a thing without being thrown overboard.

As noted, there have never been many men in any part of this country and the chances are that very few of those few have been in this gorge. If they had serious business in these parts, hunting, trapping, prospecting, or fishing, there are better places to conduct it than in the narrow canyon. Because of the difficulties of the current, travelers would tend to go around, as Franklin did through the Nine Lakes, or by another portage route. Now and then someone may have run the gorge by accident or for the hell of it, but if so they have left no signs; no old campsites, not a blazed or fallen tree. In the gorge the sense of true wilderness is tangible like that of the water, rapids, and eagles.

One afternoon Bil was working ahead, looking for a carry around a quarter mile of cascades. He came out on a wide place in the river, a little lagoon. There was a strip of gravelly beach leading back through scrub willow cover to the base of the cliff. In the brush there was a movement, a glimpse of white. He stopped, waited, watched. First one, then a second, finally five wolf pups, plump, cream colored, a few months old came tumbling down on the beach, wagging and twitching in an ecstasy of curiosity and doubt. Bil eased back out of sight, waved for the others to come very quickly but very quietly. They did and found the pups still playing on the beach. While they pups-watched, camp was set up on the far side of the lagoon. Sometime that night the pack adults who had been foraging returned, considered the situation, and apparently found it curious but not dangerous. They took the pups and retreated a bit upstream, then sat down and began to sing, making wolf music that lasted until almost full daylight.

The water in the gorge is moderately fast. There are many sections where it is possible to bounce along on the chop with only the stern paddle in the water, acting as a rudder. There are also sporty, tricky stretches of rapids and cascades, some falls that are imprudent to attempt in an open canoe, others that are impossible in any canoe. However, in terms of paddling problems—and the skill required to solve them—it is possible to find more ferocious white water within a hundred miles of New York City than in the Yellowknife. But there are different, more constant problems on this river, in this whole country, than failing to execute technical sporting manuevers. There is always danger simply because the total environment is so inhospitable. In a place where there is no help to be had and travel is so slow and difficult, a sprained ankle, an ax gash, getting lost for a few days can be a true catastrophe. It is no place to court risk for the thrill of it.

We had spent the summer with this imperative in mind. On coming to a series of serious rapids our drill was to send one canoe through first. We proceeded slowly, stopped, and took out if it seemed necessary. The paddlers above could watch this experiment, learn something of the water, from it. When the first pair came to the foot of the rapids they would turn and wait as monitors while the other two made their runs.

It was a fine half-mile stretch of white water, involving three traverses across the current, precise shots through narrow gaps in ledges, draw, sweep, pry pivots around boulders, then a final power drive through a narrow gut choked with three-foot standing waves. John and Bil thought it was the best run they had made during the summer. They shot out into a pool below the standing waves, both turned to wave at the canoes above, let each other know how well they thought they had done. As they turned, the canoe swung up on the only rock in the pool, hung crosswise to the current for a moment, then capsized.

There is a moment in such happenings of sheer astonishment. This cannot be happening. It is not possible to make a run of a half mile through good white water and then bash into the one rock in the middle of a large open pool. But it can, and when it does, questions about how and why must be immediately set aside for later review. The ice water is already beginning to burn, clutch, cramp. The canoe is rolling in the current like a sick shark, bits of gear are floating above and behind it. From above there is a yell. Terry and Ky are the best team of the group, probably the best paddlers in the Northwest Territories at that moment. Let them worry about the canoe and gear. Below the pool the rapids recommence. There is no telling how bad or easy they are or how long they run, but being caught in and rolled through them must absolutely be avoided. It is probably not smart to try to swim fifty yards directly across the pool at a right angle to the current. There is the unknown effect of the ice water. The best thing is to angle downstream, time it properly, hit the beach just before the pool empties into the rapids. It seems quite possible, but there is the behind thought, not so much panic as reality. Escaping in this manner is possible but not certain. There is a chance that all the planning, all the horse work and caution, will end in Stupidity Pool. But such speculation is also counterproductive—does not help to get a body to a beach.

There is an ultimate, kind of terminal satisfaction in knowing what you truly need and finding it; for example, a place where your foot can touch solid rock in an ice water pool. Having found what is absolutely necessary, there is no need for further wishing and speculation. Everything else, crawling out, building a fire, drying clothes and gear, emptying the canoe, going on, will follow.

Beyond the gorge there is the lower river, still fine, wild country, which we had traveled a month before but had not seen well enough because of running after ghosts. A mile or so above where it finally empties into Great Slave, the Yellowknife passes under a bridge over which runs the extension of Mackenzie Highway. As if it were a Florida causeway, there is an elderly man leaning on the bridge railing, taking the sun, fishing.

"There it is, John. The Road."

"I'd never really thought you could go beyond The Road."

"Now you know. You've been there."

Epilogue

After wintering at Fort Enterprise, Franklin and his party proceeded northward, reaching the polar sea at the Coppermine, on July 18, 1821. The Indians and some of the voyageurs turned back, while Franklin and nineteen men continued eastward for another 500 miles, sailing and paddling their canoes along the arctic shore. They reached Bathurst Inlet in September. From there, because of fall storms, they had only one choice—to march overland 250 miles to Enterprise and from there to Fort Providence. In the heart of the unexplored barren lands they were overtaken by winter. They went without real food for days on end. They were frostbitten, lost their way. They became too weak to pitch tents or cut firewood. They crawled and staggered, literally, westward. Seven of the men died of exposure and malnutrition and two of the bodies were cannibalized. A crazed Iroquois voyageur murdered Midshipman Hood and then disappeared himself.

That any of the party survived was due largely to the exertions of Midshipman George Back, who at least by modern standards may be personally the most interesting man of the group. Though he was the best artist, cartographer, most ingenuous of the Englishmen, Back rubbed Franklin and the other officers the wrong way. He was young, modish, and flip; enjoyed the company of natives, especially female ones. Yet Back was the only officer who developed wilderness wit and skills. With Franklin, Richardson, and the other survivors literally on their last legs, far too weak to travel farther, Back pushed on ahead through the winter and near the Carp Lakes found Akaitcho, secured and sent back a sledgeload of meat which saved his companions' lives. What was left of the expedition limped into Fort Providence in December 1821.

In 1825-27 Franklin, Richardson, Back were again in the high arctic. They descended the Mackenzie and then mapped 1,250 new miles of coastline between Alaska and the mouth of the Coppermine.

In the summer of 1845, then nearly sixty years old, Franklin was in command of two vessels, the *Terror* and the *Ebreus,* and 130 men who were sent out by the admiralty to solve once and for all the Northwest Passage puzzle. None of the expedition members were to return and it was not until thirteen years later as the result of a massive and continuing search operation (which did finally locate the Passage) that their fate was discovered. They had followed a freakishly open lead far into the ice pack. The lead closed and the two ships and crews remained locked in the ice for the next two and a half years. Finally, in desperation the ships were abandoned and the weakened men set out to walk across the ice to the Canadian mainland. They left their dead to the wolves and foxes along their route. Franklin himself died in June 1847. In the end the forty strongest did reach the mainland, but too late. They all died of starvation and exposure, most of them in the summer of 1848 at a place that was later named Starvation Cove.

As for the expedition of six—our fates are as yet unknown.

The Chicago Parrots

In softer, more moderate communities it might be regarded as an awful one, but in Chicago it is a better-than-average November morning. There is some high brightness behind the cloud blanket, and in consequence the urban landscape has a gun-metal sheen. It stopped sleeting about the time the invisible sun is thought to have risen. There is a skim of ice on the bean-soup-colored puddles of water in the gutters and chuckholes, but the only precipitation is a little snow coming in occasional hard, gritty flurries. The Hawk, which in southern districts of the city is what some people call the wind that stoops and rakes in from Lake Michigan, is no more than cruising. Bits of wastepaper and plastic are scattering in front of The Hawk, like plovers fleeing a peregrine, and water is being whipped against the pilings that protect Lake Shore Drive, creating plumes of spray 10 feet high. However, on the streets of Chicago The Hawk is just playing, not seriously trying to knock down the citizens.

In addition to the wind Hawk, Chicago harbors the Black Hawks, the Bulls and the Bears. But probably the most exciting and certainly the most attractive zoological gang in town is the Chicago parrots.

The parrots are a South Side outfit, and on this particular morning they are taking care of business in a 10-block area that fronts Lake Michigan between 50th and 56th Streets. The permanent center of their operations is a small park at the intersection of 53rd and the Outer Drive, where they have built a large nest 25 feet above ground in a green-ash tree. They have been working on this structure for more than three years, and it is now the size of two bales of hay and made of found twigs, grasses, shards of Styrofoam, cardboard and other scraps of urban debris.

These parrots are colonial but not, strictly speaking, communal. It seems that there are now five or six families in the flock, each with its own entrance (a hole about the diameter of a squash ball) to a private apartment in the cooperative nest. At this moment of observation five birds, using two different entrance holes, are visible on the premises, poking their heads out, strutting around a bit on the front stoops, then disappearing inside as if testing the power and inclination of The Hawk.

Hard by some tennis courts that separate the park turf from South Lake Shore Drive, eight other parrots—traveling together, as they always do—are feeding in a grove of ornamental Washington hawthorn bushes, gobbling up the frost-shriveled but still moist berries. They continue to do so for 10 minutes and then, for no reason apparent to an observer of another blood, they leave and take an aerial turn around the park, flying in tight formation, rising and dipping in a pattern reminiscent of goldfinches. As they fly they screech, very raucously. Shortly, some admiring remarks about these birds will be passed along, but this isn't intended as a puff piece on parrots. They do have their faults. The most obvious one is their voice. Their flight call is piercing, grating and unpleasant. They screech far too often in public.

By and by the birds land on a patch of parkland that is largely bare of grass by reason of heavy use by beer- and wine-drinkers and smokers of various materials. However, the place does support a good crop of hardy dandelions. The parrots start snipping off the dried fluffy heads with their heavy beaks and ingesting the pods. They move from dandelion to dandelion at a slow, staggering waddle. Without the results of an avian breathalyzer test, it's unfair to make formal charges, but they look tipsy, and it's certainly possible that they are. The thing is that hawthorn berries, like many other fruits, will, as they freeze and thaw in the autumn, ferment.The process is the same as that which enables one to make applejack by repeatedly freezing and skimming off the ice from a barrel of sweet apple cider.

Those who've kept a close watch on the Chicago parrots have noted that they sometimes get pie-eyed on hard hawthorns. They appear to be on their way to such a state this morning. Other birds—say, cedar waxwings in strong fox grapes—have this same tendency, but there is a tradition that parrots (consider all the stories about them hanging around with sailors in taverns) have a special weakness, or sophistication, in this regard. It's not all gossip. I once lived in an old house with a jungly garden high up in the Sierra Madre of southern Mexico. In the garden was a small grove of mountain coffee trees and also a magnificent parrot, a military macaw who went by the name of Siete Machos. He was a bird of such age, girth and temper that before we became acquainted he had retired from flying. It was probably a wise decision, for during the winter months he lurched about in the coffee thicket eating fallen, fermented berries. There were weeks when Siete Machos appeared not to draw a sober breath.

In any event, these eight Chicago parrots crop for about an hour on the dandelions and then rise and fly screeching back to the green ash, where they enter their private apartments in the nest. There should be another flock someplace because the most knowledgeable authorities believe there are at least 17, and perhaps as many as 25, parrots in this community. A check, however, of some of their favorite haunts— hedges around a synagogue, fire escapes on a high-rise apartment— turns up no more birds. The remainder may not have gone out this morning, preferring to wait out The Hawk in the security of the nest. Or they may be doing something somewhere nobody knows about or can even guess at, which would be in keeping with the natural history of these improbable birds.

All of which is no fanciful tease. What we have on the South Side of Chicago, in the district called Hyde Park, are absolutely real parrots. They are known formally to science as *Myiopsitta monachus* and belong to the family *psittacidae,* the parrots and cockatoolike birds, of which there are 339 species. The Chicago birds are commonly known as monk parakeets but are not to be confused with the little cage birds, budgerigars, which are often called parakeets. The monks are about 12 inches long, nearly twice as large as budgerigars; they're trim, slender birds with pointed, dovelike tails. There is a patch of gray plumage set cowl-like over the forehead (thus the name monk), but otherwise their upper parts are various shades of green, from new grass to spinach shades. The wing coverts are dark blue and the bill is rosy.

While budgerigars are from Australia, the monks are natives of South America, a numerous species, especially in the central pampas of Argentina. This, of course, raises the obvious question: What are these birds doing in Chicago? The only truthful answer is that nobody knows exactly how they got there, much less why. There are, however, theories. One frequently mentioned by bird people and in ornithological texts is that a load of monks was collected in the Argentine, circa 1967, and shipped to the U.S. to be sold as pets. At Kennedy Airport in New York a crate of birds was dropped and broke open, the captives escaping. Not long afterward, a number of monks were seen in the New York City area. During the next few years, individually or in small groups, monks were reported in Asheville, N.C.; Plymouth, Ind.; Muskegon, Mich.; Dallas; Norman, Okla.; Omaha; and Anaheim, Calif.

However, as serious birders will admit when pressed, the idea that adventuresome parrots, splitting from JFK, flew purposefully westward, passing over the orchards of Pennsylvania, attractive parks in Cleveland and the cornfields of Indiana, to settle in the Hyde Park district of Chicago, goes a long way beyond belief.

Dr. William J. Beecher, the Director Emeritus of the Chicago Academy of Sciences and a distinguished ornithologist, has given considerable thought to the matter. He says, "The Kennedy story as explanation for the birds in Hyde Park isn't plausible. It's contrary to everything we know about how imported species, as for example the starling [which was introduced to this country in the 1890s and about which Beecher is a world authority], become naturalized and expand their range. Such species disperse in a continuous pattern, establishing themselves as breeders in suitable habitats and then pioneering farther. They don't hopscotch, turn up in widely separated population pockets as the Kennedy theory suggests the monk parakeets have done."

Doug Anderson is a resident of the Hyde Park area and a passionate, well-regarded amateur birder who is a past president of the Chicago Ornithological Society and currently the vice-president of the Chicago chapter of the National Audubon Society. He has watched the 53rd Street parrots longer and more closely than any other expert. Anderson says he has repeated the JFK escape story himself because it has become more or less the official one, and he believes there may be a connection between the airport escapees and the birds nesting in Hyde Park. Both he and Beecher, however, point to the fact that breeding populations in the 1,000 miles between New York and Chicago are, if existent, rare and isolated.

Anderson feels it is also possible that the Chicago parrots escaped directly from pet suppliers or private owners in the city, or were intentionally released. This sort of thing, as hundreds of stray dogs and cats, occasional white rats, once-tame raccoons and alligators attest, is a common occurrence in urban areas. (Last summer, a 7-foot boa constrictor was roaming freely about western Chicago, and a fisherman snagged a piranha out of the artificial lagoon in a municipal park. The former was accidentally run over and killed in a shopping-center parking lot, and the latter, after being caught, was sent to the Shedd Aquarium, where it was dissected for purposes of accurate identification.) Certainly, it's conceivable that having come by a pair of monks and having sat around listening to them screech in, say, a small apartment, a disenchanted birdkeeper might open the cage door and shoo them out of the nearest window into the world. Even so, the local escapee explanation has its weaknesses: Monks have been common in the pet trade for decades. Why would they suddenly and only recently show up in Hyde Park? Also, why did they appear as a flock? Was there a mass release, or did birds that had gained their freedom individually or in pairs flap around the city until they found each other?

There is another theoretical possibility: that for inscrutable reasons, a crowd of monks left the Argentine, started flying north and didn't stop until they reached 53rd Street, which they found especially attractive. This is not altogether fantastic; smaller and less robust birds travel twice a year between the very northerly and southerly parts of the hemisphere. Nevertheless, the idea of the monks making a mighty one-way emigration is so much at odds with current zoological knowledge that it is virtually unmentionable in serious bird circles. Also, it's impolite, even cynical, for anyone outside those circles to suggest that theory because, like the one concerning the Kennedy escape, it can't be factually disproved, and the only answer can be, "Everybody knows that something like that can't happen."

However and from wherever they got there, the monks first appeared at their present location in Hyde Park in the spring of 1980. They have been in continuous residence ever since and have raised at least five, and perhaps a dozen or more, young birds in the green-ash nest. As far as most experts now know, they are the most northerly breeding colony of their kind to be found anywhere in the world. A question that has occurred to many on first seeing or hearing about these birds is, How have they survived the harsh Chicago winters, especially the one of 1981–82, the most ferocious the area has had in decades? The answer seems to be, rather easily—better, in fact, than many other creatures, including people. In the worst part of 1982 Beecher found the frozen carcasses of two monks, but he cautions that other factors, injury, for example, may have been the primary cause of death. Beecher feels that having made it through three winters the Chicago birds have demonstrated that climate alone isn't a bar to them. This isn't particularly surprising, since in the southern parts of their natural range in Argentina, winter weather is brisk, if not quite as frigid as it is along the Lake Michigan shoreline.

Anderson reports that the monks have made at least one adaptation to the climate. In the worst weather, when The Hawk is raging, the birds will temporarily abandon their nest to seek shelter provided by buildings, which are more effective windbreaks than their own twig-and-grass structure. Among the most popular refuges are the fire escapes on a large South Shore Drive apartment building, which is fittingly, in this context, called The Flamingo.

For many animals the scarcity of food during the winter is more crucial than low temperatures and is the reason they migrate or hibernate or, if they don't, sometimes perish. The Chicago parrots have coped reasonably well. From early spring until late fall, local parks and ornamental plantings around nearby buildings seem to provide plenty of food of the hawthorn-dandelion sort. Also, there are several residents who enjoy feeding pigeons on neighborhood streets. The monks have learned about this and congregate for those handouts. In the dead of winter, the time when natural foods have disappeared or are covered by snow and ice, the parrots visit—and probably survive because of—private bird-feeding stations, which are surprisingly numerous in Hyde Park.

David and Sylvia Smith and Bob and Rita Picken are next-door neighbors on 56th Street, and both couples have feeders in their backyards. Late in 1980 Sylvia was the first to see one of the parrots. She was astonished. "It was fairly early in the morning and I got David out of bed because I knew unless he saw it himself, he wasn't going to believe we had a parrot on our feeder," she says.

Since then the Smiths and Pickens have had parrots regularly during the winter, often as many as 17 around a feeder at the same time. Both couples are pleased, enjoying the exotic birds for their own sake and also for the incredulity of those who haven't previously known about them. Sylvia Smith says, "We have people stop and knock on the door to tell us they think we have parrots flying around the yard. A lot of them seem to think we should do something about it; catch the birds and put them in cages, or at least report it to the authorities. We say, 'Oh yes, those parrots live around here. We just watch them.'"

When the parrots arrive at feeders (some are window-ledge trays outside 18th floor apartments), they monopolize them. Because of their numbers and pushy temperament, they drive away even blue jays and cardinals, which also are found in the neighborhood and are generally aggressive creatures themselves. As a rule, the parrots don't leave until they have cleaned off the trays. They scarf up virtually any kind of commercial birdseed. "They are like little vacuum cleaners," Rita Picken says. "They scoop up everything. I've never seen any birds that eat as much and as fast as the parrots do."

Understandably, the parrots are a lively conversation piece, and exchanging thoughts about them draws even strangers together. For example, while watching the birds one morning I struck up a friendly conversation with a fellow who had so much the physique and manner of a treetop lover that I thought he might be LeRoy Brown. He introduced himself only as Top Hat.

Top Hat lives a considerable distance south of Hyde Park. He said that he hadn't previously heard about the parrots and as a rule wasn't much interested in birds. However, having come to 53rd Street on other business and seeing the parrots, he wasn't ashamed to admit he was enchanted by them. We talked for a time about the usual things—where they came from, what they ate, how they survived the cold. Then Top Hat raised a good question: "Now you tell me why all them regular birds, you know the ones that belong here, why don't they all get together and kick them little green mothers' asses out of here?"

Anderson, the Hyde Park ornithological expert (he has identified 235 species in the immediate area), phrases it differently but has been thinking about the same thing—competition between the monks and the longer-established species. Anderson makes a number of interesting points. The first is that of the 339 parrot species found in the world, only the monks build nests in the open. The other 338 occupy, and lay and hatch their eggs in, found cavities, usually holes in trees. Beyond the fact that this may make the monks more adaptable to new territories than most parrots, it strongly recommends them to a concerned birder like Anderson. Ornithologists generally take a dim view of imported—"foreign"—birds becoming established in this country, in part because of what happened with the starling. Starlings are aggressive, fertile creatures and also hole-nesters. As soon as they got their feet on the ground in this country, so to speak, they began taking over the nest cavities of bluebirds, woodpeckers and other birds, sometimes literally kicking out the previous residents. The success of those pugnacious imports is thought to have brought about the decline of some species that weren't able to compete.

The parrots, at least, won't cause a starling-type problem, nor does it seem that their nesting habits will in any way inconvenience established birds. No other local species is colonial, as the monks are, and even in a city there is always an excess of tree space for avian homebuilders. Also, the monks won't annoy their human neighbors as do pigeons and house sparrows, which often construct sloppy, shanty-type nests on window ledges and under eaves and cornices.

Anderson has noted only one interspecific consequence of the parrots' building their nest in the green ash on 53rd Street. Occasionally, gray squirrels, plentiful in the park, have been observed sitting, running about and perhaps catching food on top of the bulky parrot quarters. It's quite possible that once they become more familiar with the facility, a squirrel or two may burrow in and take penthouse quarters above the birds. Anderson says the parrots seem aware of the squirrels but don't appear especially alarmed or angered by them. This phlegmatic response may be something of a conditioned one. In South America, a small species of opossum frequently has been found denning up in vacant parts of apartment nests built by the monks.

Except for being greedy and a bit overbearing around the neighborhood bird feeders, the Chicago parrots, at least in their present numbers, have not, by competing for food or anything else, created hardships for any other creatures. "They seem to be filling a niche [both a place and ecological role] that presently isn't occupied by another species," says Anderson. "I suppose in a sense they may have moved into the niche that once was occupied by parrots here."

The niche Anderson is referring to was filled by the Carolina parakeet, which until it became extinct in 1914, was the only parrotlike bird found in most of the country when European settlers arrived. (There is a species, *Rhynchopsitta pachyrhyncha,* the thick-billed parrot, which occurs in a few mountain areas of southern Arizona and New Mexico.) When both species were extant, the Carolina and monk parakeets were relatively similar, both belonging to a subdivision of small New World parrots called Conures. They were both principally fruit- and seed-eaters and traveled in large flocks. In the 17th and 18th centuries the Carolina parakeet population was enormous. Despite its name, the species ranged throughout the East, from the deep South to the Great Lakes, and was once prevalent in the area that has become Chicago.

Indisputably, we are responsible for the complete demise of the Carolina parakeet. A few were kept as pets, some were hunted for the pot, ending up in game pies, but in the main they were exterminated by farmers and orchardists who dislike their feeding on grains and fruits. The destruction of the species was hastened by an odd habit of the birds. By report, they were never especially shy, and a man with a gun could approach a feeding flock. After the first blast had felled some of the little parrots, the rest would fly only a short distance away, then return, individual birds lighting down by their dead and dying companions to brood, figuratively, about their condition. They would repeat this behavior "until," wrote a chronicler of these expeditions, "sometimes an entire company would be wiped out."

"I know it is a bit iconoclastic, given the feeling about imported species," says Anderson, "but I enjoy seeing those birds in Hyde Park, and I hope they become established. I never saw a Carolina parakeet. I like the idea of the parakeet niche being reoccupied in this country."

As Anderson infers, not everyone has been pleased about the coming of monk parakeets to the U.S. In fact, what might be called the official reception of the species was on the hostile-to-violent side. In the 1960s, after the first feral monks began to appear—from Kennedy Airport or wherever—the U.S. Department of Agriculture sent an all-points warning about them to appropriate federal and state wildlife agencies and to some private organizations such as the National Audubon Society. The gist of the alarm was that these were potentially bad birds and that it would be a righteous act if those finding them, or particularly their nests, were to destroy them. The agricultural knock against the monks is that in South America they are known to eat a lot of corn, millet, sunflowers, grapes, apples, peaches and other fruits and grains. Flocks have been known to destroy as much as 15% of some crops in certain areas. In attempting to control the birds in the Argentine, farmers and government agents have had at them with guns, traps, fire and poison. In one two-year campaign, bounties were paid for nearly half a million dead parakeets; two legs of a monk brought the peso equivalent of one cent. Nevertheless, the population remains vigorous throughout most of its range.

Whether or not the monks could become numerous enough in this country to cause serious crop losses isn't known. Since feral monks that have been reported regularly during the last 15 years have always appeared in very small groups, it's possible that they may have the capacity to be only a marginal breeding species here. However, on the chance that it might be otherwise, the Department of Agriculture issued its warning.

The fears of agriculturists and their public servants about parrots may or may not be well-founded; at least they rise directly out of practical self-interest and deserve to be considered seriously when it comes to dealing with the species. But another reason for the cold reception the monks received in some quarters is an almost reflexive bias against species that have been brought to this country recently and which then start living freely here. In support of this, the trouble supposedly caused by imported starlings is often cited. (In fact, in many areas starlings are now the most interesting and ecologically significant elements in the avian community.) However, this judgment is essentially an esthetic one, having less to do with science than with zoological snobbery and chauvinism.

Much of this country is occupied and operated by immigrants, that is, species that didn't originate here. This is most apparent among ourselves, every person jack of us—red, white, black, yellow and mixtures thereof—having ancestors who came here from other continents. The same is true of a good many of the resident plants and animals. In thinking about this, we tend to make arbitrary judgments, based largely on time and manner of arrival, about good natives and undesirable aliens. The Everglades kite and black-footed ferret, which got here many centuries ago by their own devices (from the Caribbean and Siberia, respectively), are deemed to be fine, natural native Americans, altogether deserving of our support and sympathy. Another all-right category includes such things as the honeybee. It was imported, but so long ago—in the 17th century by English settlers—that it's now celebrated as a good old American bug. On the other hand, something like the monk parakeet is regarded with scorn and suspicion, as an impure, essentially not-nice creature, because it came (maybe) in the last 20 years by (perhaps) jet.

Underlying all this is a more pervasive and mischievous conceit: that we humans are the unnatural species. There are two versions of this notion. The first, chronologically and probably still in terms of popularity, is that we are unnatural because we are so far above and better than other living things. The second, which has gained ground rapidly in this age of exquisite environmental sensibility, is that we are unnatural because we are worse than everything else. The corollary of this is that other phenomena which are unduly affected by us and our works—say golf courses, Toggenburg goats, park pigeons and zoo chimpanzees—are no longer really and truly natural. These notions are supported by little but hubris and/or a kind of Peter Pan–Captain Hook romanticism. All the hard evidence points to the conclusion that we belong here and are as natural as anything else. In fact, at the moment, because of our numbers and energy, we are a force of nature on the order of fire, frost or flood. Some may prefer the Tribune Tower in Chicago to El Capitan in Yosemite, or vice versa, but both were created by natural powers.

Our impact, like that of sun or wind, seems, at least in the short run, to be sometimes good and sometimes bad. We've played terminal hell with the passenger pigeon and the ivory-billed woodpecker. We have seriously discommoded the whooping crane, wolf and grizzly, the American chestnut and the elm. On the other hand, a lot of things—blackberries, locusts, white-tailed deer, crows, possums, skunks, woodchucks—are more numerous and vigorous because our activities during the past few centuries have improved the environment from their standpoint. Also, we have recently proved to be major dispersing agents, serving to greatly diversify the flora and fauna of given areas. If nature is thought of as what *is*, rather than what once was or should be, then it must be acknowledged that ours is much more various than it was 300 years ago. There are hundreds of species of plants and animals—herds of Herefords, flocks of pheasants, plantations of perennial flowers—that are now thriving here because of us. (As a matter of statistical record, there are certainly 39 species of birds, and probably 56 that were imported to North America by design or accident and have become naturalized. From this increase must be subtracted the six species we have exterminated.) In the process, we have done some dumb, cruel, malicious and wantonly selfish things that now do not seem to have been in our own or anybody else's best ecological interest. But we have committed no unnatural acts.

This and more bring the parrots back to mind. A 707, or whatever they arrived on, is a no more despicable form of transport than the southwest wind, which some centuries ago carried the Everglades kite to these parts. Somewhere down the road, the monks may become established here. If so, we may regard them as excellent additions; bright, pretty, useful birds that help to make up for the loss of their kin, the Carolina parakeet. On the other hand, they may turn into awful pests, making us wish they had stayed home to harass the Argentines. But that is nature for you. The most beautiful and comforting thing about it is that it is ultimately so complex as to defy even medium-range prophecy and too powerful to be controlled by any of us.

Now, as the parrots face their fourth Chicago winter, the colony is thought to be the largest established group of monks in the U.S. The location is serendipitous for the birds. The little park on 53rd Street where the nest tree stands occupies only a single block, but as genuine natural history goes, it is a fascinating bit of habitat. In times past it lay under the waters of Lake Michigan, then was a strand of dune and more recently prairie. Now it is about as heterogeneous in terms of life forms as any few hundred square yards on the face of the earth. In addition to the parrots, the most common birds in the park are pigeons (actually rock doves, natives of Asia Minor that were brought at some forgotten date to the British Isles and transported from there to here in English ships in 1621); house, or English, sparrows that originally ranged from the Atlantic coast of Europe to Lake Baikal in Siberia and were intentionally brought to and released in the U.S. in 1850 in hopes that they would help clean up horse manure from the streets of eastern cities; and the aforementioned starlings, which evolved in the East Indies, spread or were spread across Eurasia before being imported to this country as part of a romantically conceived plan to introduce all of the non-native bird species mentioned in the works of William Shakespeare. Growing around and in the park in Chicago there are hawthorns of a species first found by us in our South Atlantic regions, Chinese chrysanthemums, English ivies, French marigolds, euonymous, forsythia, the yews, privets and dandelions, along with the oaks, maples, ashes and other flora of Illinois. As for the human community, facing the park are the Hampton House and Del Prado apartment buildings, a synagogue, Congregation Rodfei Zedek, the Pierre Andre Hair Styling Salon and the House of Eng, a Thai restaurant. In the park is a brick comfort station. Its outer walls are inscribed with wonderfully exotic graffiti: ELIJA IS DEAD MALACHI 3-22-24. WATCH POLAND. WHO ARE THE JEWS. I KNOW NO PEOPLE OF EDOM THAT WHO THE JEW ARE.

The more you hang around this micro-niche the more ordinary the presence of green parrots from Argentina seems. There are all sorts of creatures that came as far to reach this park, and probably by means that were at least as unusual—beings within whom there are elements, if not memories, of far Cathay and Krakow, the South Downs and great steppes, the Spice Islands and Baltic forests, jungles and deserts, Seville and Samarkand. All along 53rd Street are epics and myths in waiting, tales and histories that would not suffer in comparison with the Chronicles of Narnia and the Shire. One might commence: "Sore pressed were the Green Screechers of the pampas. Beset by fell gauchos and game protectors, rent by shot, choked with foul fumes, their ancient high homes enflamed, the Company of Wanderers gathered and took wing. Oared they stoutly the air toward Farnoth, the great lake of frost and The Hawk, where it is said now lies their hopes and fates." Fear of anthropomorphism be damned!

Ordinary observation suggests that in terms of true ecology—in which human forces are properly weighed—the Hyde Park neighborhood is one of the most benign imaginable habitats for monk parakeets available in this country. By contrast, had they tried to settle in a national park, state game refuge, private sanctuary or other place managed according to fashionable and up-to-date ecological principles, they would almost certainly have been rousted. I have a vivid recollection of a federal naturalist explaining why he was going to cut down a Norfolk Island pine tree that once grew in the Everglades Park. The tree looked to be about 75 years old and was a fine specimen that provided shade, food and shelter for a diverse community of local plants and animals. The ranger said it was a species native to Australia, planted by some environmental ignoramus. It had to be removed to protect the natural purity and integrity of the park.

Most Hyde Park residents aren't that sophisticated when it comes to this sort of high environmental seriosity. (Anderson is an exception, but with his ideas about the monks filling the niche we emptied of Carolina parakeets, he is something of a rare bird himself.) The people who live in the area seem to be glad the parrots are there and don't bad-mouth them because they are recent and irregular immigrants. The apparently unanimous feeling is that the birds are pretty to look at, interesting to talk about and a desirable neighborhood resource that should be protected.

At least twice during the last three years, outsiders have tried to molest the parrot nest. Once a fellow threw a rope over a lower limb on the ash tree and started to climb toward the nest. On a second occasion an intruder hung around the foot of the tree with a long-handled net, trying to catch the birds. The assumption is that these raiders were motivated by greed, that they thought they could sell the birds for big money. This isn't the case, since a monk retails for only $25 or so in pet shops. Whatever their intentions, these strangers were frustrated by local residents who saw what they were up to and drove them away.

The entrance to the Hampton House apartments is directly across 53rd Street, only about 100 feet from the parrots' nest. In consequence, Ronald Faulkner, the front-desk man, is, day in and day out, closer to the birds than anyone else in the neighborhood. He says that being able to see so much of the parrots is a fringe benefit of the job, and "I tell you this: Nobody better mess around with them parrots. I'm watching, and we have a lot of senior citizens around here. They care a lot about the parrots, and the pigeons and the squirrels in the park, too. They may be old, but they'll make some big trouble in a hurry for anyone who tries to do anything bad to those animals."

Because of recent political developments in Chicago, the informal vigilante force that keeps a protective eye on the parrots has been heavily and officially reinforced. Now the nest area is in effect a maximum-security sanctuary. This is because a resident of Hampton House is Harold Washington, who last year, in a campaign of notable bitterness, even for Chicago, was elected mayor of the city. Since then at least two armed and uniformed police officers have been on duty at Hampton House, either posted at the door or in a squad car, which is usually in the park, directly under the nest tree. The officers are primarily responsible for the welfare of the mayor, but anyone with evil designs on the parrots would now literally have to shove the cops aside to get at the birds. Their own feelings aside, the police wouldn't be likely to permit this, because His Honor has made it clear that he thinks extremely well of the birds; that catching a glimpse of them in the morning as he leaves Hampton House has brightened up some otherwise hard days, of which he has recently had more than his share. Indeed, he gives a ringing, pro-parrot tribute: "We are all pleased and grateful that these fine parrots have chosen to settle in the great city of

Chicago. I think of them as an omen signifying better times ahead for the entire community. For me personally, they have been a good-luck talisman. On the South Side we all admire them for the way they face and stand up to The Hawk."

Sheer serendipity is the only plausible explanation of why the mayor and parrots of Chicago are next-door neighbors. However, this is a neighborhood where people are not afraid to play around with deeper realities. For example, a young man well under the voting age, in fact a fourth-grade student, while cutting across the park one afternoon, stops to talk parrots and, in a sense, politics. His name is Jerome. Jerome says, "Those green birds was brought here by the mayor. The mayor lives in that big building right by the birds. They are his good luck and they are lucky for everybody." Brian Boyer, a press aide to Mayor Washington, says that Jerome's is only a creative embellishment of an opinion or hunch that he had heard expressed previously. "There are people who know about the mayor who don't know about the parrots," he says. "But nearly everybody who knows about the parrots—and a lot of people on the South Side do—knows they live across from the mayor. If people talk about the parrots, they almost always say something about the mayor. He's had pretty good luck, and so people make a connection that the parrots, like he says, are good omens, like finding a penny or a comet or something. Maybe that's how folk stories or myths get started."

Given the ecological, social and now political realities of their present habitat, it seems most unlikely that agricultural agents, wildlife managers or earnest environmental purists are going to have much influence on the Chicago parrots. Whatever it is that brought them to this safe haven—chance, zoological complexities beyond comprehension or mysterious higher authorities—theirs, most assuredly, is a thought-provoking and natural history.

The Curiosity of Thomas Nuttall

Wilson Hunt was a St. Louis merchant and an associate of John Jacob Astor, the principal organizer and beneficiary of the American fur trade. In 1810 Astor picked Hunt to lead an overland expedition of furmen to the Pacific. Coming as it did just four years after the return of Lewis and Clark, it was to be the first privately sponsored venture of this sort. The plan was that Hunt was to lead his men across the mountains to the mouth of the Columbia River and there meet a sea-going section of the party which would have sailed from New York and around Cape Horn. The combined force would build a fort from which Astor's agents could challenge the British-backed Hudson's Bay and North West fur companies. It did not work that way. The post was not established, and before the expedition ended, forty-seven of the ninety-eight men in the combined party died. Another went mad, and the remainder suffered terribly before they straggled back to the States.

These disasters of course lay unsuspected in the future when the overland Astorians left St. Louis, traveling up the Missouri in the early spring of 1811. Hunt commanded the largest, best-supplied fur brigade to go upriver that year, and he was full of himself and optimism about the project. Among others, he was accompanied by two itinerant naturalists whom he had met by chance and invited to go along, at least as far as the upper Missouri. That the party should start out with two such supernumeraries is not surprising, given the temper of the times. Hunt may have proven an inept frontiersman, but he was an educated eastern gentleman, influenced by contemporary cultural currents of thought and fashion. One such was a growing fascination with the natural history of the North American continent. Satisfying the curiosity of both the intellectual community and the public at large about what actually grew, stalked, and lay about in the western hinterlands had even become something of a matter of public policy. Among others, expeditions led by Meriwether Lewis, William Clark, Stephen Harriman Long, John Charles Frémont, while semi-military in organization, had been or would be charged to collect and bring back natural history information about such things as (as Thomas Jefferson had instructed Lewis and Clark) "the soils and face of the country; the growth and vegetable productions; the animals of the country and especially those not known in the United States."

Government probes and probers aside, the period between the founding of the republic and its Civil War was preeminently that of the great freelance naturalist-catalogers; of André and François Michaux, William Bartram, Alexander Wilson, John James Audubon, and Thomas Say. It was a grand time for such endeavors because there were so many new discoveries to be made. Fortunately located, a keen observer could bag an undescribed bird before breakfast, an odd geological formation and a shrew by lunchtime, and add half a dozen new plants to the catalog of science between then and dark. Fine points of taxonomy, speculation about ecological relationships could and did wait. The obvious need was to find and approximately identify what was out there. With a high sense of purpose and passion, the field men set out to make the enormous inventory, collecting, pressing, pickling, making notes and sketches about what they found. They were likely to turn up almost anywhere, at a trappers' camp or in a backwoods clearing, often ragged and broke, not uncommonly cold, wet, hungry, and lost. Even the untutored generally received and regarded them well, treating them as they might a wandering preacher. Sometimes they were twitted about their apparent lack of what was regarded as frontier common sense, but their persistence in the face of formidable difficulties and hardships often won admiration. The consensus seemed to have been that the work of these "perfessers" was a Very Good Thing, deserving of full moral support and such material assistance as could be conveniently spared.

Under the circumstances it is not unusual that Wilson Hunt met up with two naturalists or that he invited them to travel along with his expedition. At very little expense—two additional rounds of rations and occasional transport on a horse or in a flatboat—Hunt had the satisfaction of becoming an instant patron of the popular arts and sciences. For the two naturalists, a John Bradbury and Thomas Nuttall, the patronage was essential. It allowed them to live without expense while traveling westward, and both were poor men. More importantly, it was then virtually impossible to travel very far beyond the Missouri settlements without the protection and support of either an armed fur trading party, such as the Astorians, or the military.

Both men were English, which was not surprising since British intellectuals were at least as interested in the natural history of North America as their American counterparts. Both were generalists with a lively interest in all manner of phenomena but with special training in botany. Each had nominal backing from eastern individuals and institutions. Bradbury, the older of the two, had been commissioned by the Liverpool Botanic Garden to bring back living plants and seeds; bird and mammal skins for an affluent private collector. Nuttall, who turned twenty-five a few months before the Astorians left St. Louis, was under a peculiar contract (of which more shortly) to investigate virtually every known branch of natural science for a Dr. Benjamin Barton of the American Philosophical Society, then perhaps the most important scientific body in the United States.

By separate routes the two Englishmen had arrived in St. Louis, Nuttall after a considerable wilderness trek through the northern Appalachians and Great Lakes area. Being the only professional naturalists in the immediate vicinity, they had become acquainted, collected and presumably philosophized together. However, they were not close friends, nationality and science aside, being quite different men. Bradbury by all accounts was reasonably convivial, something of a hunter and outdoorsman who fit in well with trappers and traders and in fact had the skills and outlook which, under different circumstances, might have made him one of them. Nuttall emphatically could not be mistaken for a frontiersman. He could not handle a gun proficiently, paddle a canoe, make a decent camp, sit a horse properly, or swim at all. That he had survived as long as he had in the wilderness seemed to some of his contemporaries to be evidence that there was a force somewhere which took a protective interest in the innocent and unwary.

Nuttall was a small man already balding with, according to acquaintances (he himself was above self-portraiture), a "cerebral" but strangely childlike face. Except in the occasional company of other naturalists, he tended to be shy to morose. Certainly he had absolutely no inclination or talent for sitting around a camp with the Astorians singing bawdy songs or drinking raw whiskey. More than just not fitting in with the rough and ready company, he apparently was oblivious to it and the practical realities of the expedition. Shortly, strange stories began to circulate about the peculiar behavior of the little botanist. Some were very likely apocryphal, frontier stretchers, but there is an underlying sense that the furmen soon decided that they had a very strange 'un on their hands, not a plausible man like John Bradbury doing implausible work, but a truly odd duck doing odd work.

During the first weeks the party moved up the Missouri in flatboats. For hours or sometimes days on end, Nuttall would leave the flotilla to roam through the bottomlands and prairies. Often he forgot or could not bother to take food, bedding, and weapons with him. Not uncommonly he became so engrossed in his sciences that he did not know where he was in relation to the boat party and the Astorians would fetch him back. Occasionally he was brought in by wandering Indians who came to the conclusion that he was a powerful white shaman, nothing else seeming to explain his rash activities. Now and then, to fortify themselves on such rescue missions, the tribesmen would filch and drink the raw alcohol from Nuttall's collecting jars.

In what is now South Dakota, the Astorians had their first serious trouble with the Teton Sioux, the formidable tartars of that time and place. The Tetons were not at all pleased at the prospect of the whites trading with and bringing guns to lesser tribes. To emphasize the point some six hundred warriors, perhaps the best light cavalrymen the world was ever to see, drew themselves up and blocked the passage of the Astorians. The commotion was sufficient to catch even Nuttall's attention. Dutifully he shouldered his musket and lined up with his fellows, ready to do what he could to help repulse the hostiles. However, before it came to blood, the Sioux were placated by negotiation and bribes, and the fight was avoided. This was especially fortunate for Nuttall. It was discovered that the barrel of his gun was tightly packed with packets of seeds which he had stuffed there to keep them dry. Had he fired the rickety piece (which he had also used as a digging bar to uproot flora specimens), it would likely have exploded and, even before the Teton Sioux got in their licks, probably deprived America of its most extraordinary nineteenth-century naturalist. The incident became something of a western myth, told and retold on trails and around campfires when the subject turned to absent-minded perfessers.

Traveling upstream in the summer of 1811 with another fur party was a third supernumerary, Henry Marie Brackenridge, a young man of good family and Princeton University connections. He spoke of himself as a geographer but now might be called an environmental journalist, being generally interested in the lay of the new lands and in preparing some articles about them for the eastern press. In time he became acquainted with Bradbury and Nuttall and later struck off an interesting biographical paragraph about the latter. Mentioning Bradbury first, Brackenridge then wrote:

"Mr. Nuttall, engaged in similar pursuits to which he appears singularly devoted, and which seem to engross every thought to the total disregard of his own personal safety, and sometimes to the inconvenience of the party he accompanies. To the ignorant Canadian boatmen, who are unable to appreciate the science, he affords a subject of merriment; *le fou* is the name by which he is commonly known. He is a young man of genius, and very considerable acquirements, but is too much devoted to his favorite pursuit and seems to think that no other study deserves the attention of a man of sense. I hope, should this meet his eye, it will give no offence; for these things often constituted a subject of merriment to both of us."

It was the first of many similar comments. All naturalists were expected to be curious about the phenomena they were scrutinizing, and because of their dedication to such an impractical calling, there was a tendency to regard most professional naturalists as curious—that is, somewhat peculiar fellows. However, as his American reputation grew, many others were to come to see Nuttall as Henry Brackenridge first did in that summer of 1811—as the most curious, in both senses of the word, of the whole lot.

As to his idiosyncrasies, real or rumored, some found them charming and were attracted to him not only because of his knowledge and insight but by what seemed to be a peculiar simplicity and innocence. John James Audubon was one of those. When, in 1832, he first met Nuttall (then by far the more famous naturalist), Audubon was enthusiastic, writing a friend, "Nuttall is a gem—quite after our heart, and I am very happy to know him as such."

Others were less enchanted by some of Nuttall's traits, finding his absent-mindedness exasperating; judging him to be excessively reclusive, something of a miser and misanthrope, occasionally a sponge and a sloven. Virtually all of these diverse assessments and in fact nearly all of Nuttall's social relationships were in one way or another connected with his passion for natural history. Those who could appreciate it, Audubon, for example, found him agreeable, outgoing, and generous. Those who did not were, it seems, largely ignored as non-persons by Nutall, while he dismissed those who crossed or frustrated his interests as fools or worse. One of the few reports of Nuttall in a tantrum occurred years later when he had become a towering figure of American science. Members of the Philadelphia Academy of Natural Sciences, not unreasonably, asked that he tidy up the piles of dried plants which were littering the common rooms of that institution. Enraged, Nuttall remarked that "for any interest taken here [at the academy, which he described as a "would-be" scientific society] in botany, one might as well be amongst the Hottentots themselves."

Opinions about what Brackenridge and many later acquaintances would see as Nuttall's genius were more consistent. By the time he was in his forties he had published the first and for many years best and most comprehensive reference books in America about the plants and birds of the continent. Without much dispute, he was the foremost botanist and one of the leading zoologists and mineralogists of his time. More impressive even than his publications was his field experience. He was to pioneer into more corners of the land, previously unknown to naturalists, to see for the first time more new species and wonders than anyone had before or would thereafter. In consequence his name became and remains indelibly associated with hundreds of plants and animals, either as the man who first discovered and described them or as the man for whom they are named—Nuttall's dogwood, Nuttall's poorwill, Nuttall's cottontail, Nuttall's cockle, and many more. Quite simply and literally, Thomas Nuttall is an immortal of North American natural history and will remain so as long as anyone takes a systematic interest in the subject.

Despite the unique record of his accomplishments, his own considerable writings and those of others about him, the portrait of Nuttall, the man, which survives seems at first glance to be flat and incomplete, a caricature of the eccentric genius such as first made by Henry Brackenridge. Thus we have voluminous notes on how Nuttall stood in regard to, say, taxonomical issues of his day, but very little about his political, religious, economic opinions; almost nothing about what he did and was when he was not being a naturalist—about his nonprofessional friends, amusements, ambitions, and agonies. Perhaps there was a private side which he kept well hidden or that nobody noticed and has been forgotten. However, this seems unlikely since he became a prominent figure in the small but ferociously articulate intellectual–academic community of the early nineteenth century. Had he been an opera-goer, a Swedenborgian, a whist player, a secret womanizer or drinker, it is inconceivable that such activities would not have been ferreted out, gossiped and inevitably written about by contemporaries.

The best explanation may be that there was no hidden man and that he was exactly what he seemed to be. Very likely there is now and was then little information about the non-naturalist side because it was so insignificant, especially to him. If so, it would seem to make him a more rather than less interesting figure. Beyond the mighty record of accomplishment, Nuttall was perhaps one of the most extraordinary personalities of his day, or any other, by reason of his remarkable, single-minded preoccupation with the natural history of North America. At the very least, such a theory—passion to the point of monomania—better than any other seems to acount for how and why this small, reclusive, badly prepared and equipped Englishman got himself to the upper Missouri River in 1811.

In 1800, the fourteen-year-old son of a widowed mother of small means but good family connections, Nuttall left his native Yorkshire and went to Liverpool as an apprentice to an uncle who had prospered in the printing business. During the next seven years he learned the trade (well enough so that he occasionally practiced it in North America in order to keep body and soul together) but more importantly somehow became enthralled with, as he said, "The Goddess Flora." What patron, friend, garden, book, first directed Nuttall to his botanical passion is one of those personal details he felt to be too trivial for explanation. All that is really known is that in 1807 he told his uncle he was finished as a printer and was going to the United States to study and admire its flora. The uncle, Jonas, a childless man who had dynastic ambitions for his nephew, was astounded and angered by the proposal, apparently feeling that the United States was an improbable place at best and that hunting flowers in it was as preposterous as retiring to a garret and writing poetry. Whatever Jonas' arguments, they had absolutely no deterrent effect on young Thomas, who sailed from England in 1808 and arrived in Philadelphia on April 23rd. There, without any hesitation or doubt, he immediately set about doing what he came to do and would continue to do for the next thirty-five years— that is, naturalize. Without stopping even to find a room, he put down his baggage and walked out High Street, crossed the Schuylkill River, and began looking at plants—the greenbrier being one on that very first day which particularly caught his fancy and set him speculating about its proper taxonomical position in the botanical scheme of things.

Thereafter he did find lodgings of a sort and got a part-time job in a print shop which enabled him to live and at the same time to start traveling up and down the Middle Atlantic area, looking for and at its flora and fauna. Also, Nuttall began to make the acquaintance of members of the Philadelphia intellectual set, then, especially in regard to natural history, the most prominent in the United States. Among others, Nuttall was eventually introduced to Dr. Benjamin Smith Barton, officially one of the great figures of the scientific community. A professor of medicine at the University of Pennsylvania, vice-president of the American Philosophical Society, affluent and well connected, Barton had his fingers in a great number of scientific and intellectual pies. However, his own work tended to be thin and pretentious, and in fact he was more a cultural politician than a scientist.

Among other things, Barton had published a botany and had the large ambition to author a comprehensive reference dealing with

American flora. However, he had neither the time nor talent to do the hard collecting and classifying work that such a project entailed. Therefore he had taken to engaging promising young men to do the research which he then hoped to use in and claim credit for in his masterwork. Nuttall obviously struck Barton as an ideal candidate for this sort of thing, being dedicated, industrious, and so green in the ways of scientific infighting as not to give any troubles about authorship or recognition. In 1810 Barton made Nuttall a formal offer of employment, drew up a contract which remains one of the extraordinary documents in the annals of either science or labor relations.

Barton proposed that Nuttall travel across Pennsylvania, then north to the Great Lakes, and from there on into Canada, pushing up to the subarctic regions in the vicinity of Lake Winnipeg. Thereafter he was to work south and west to the upper Missouri ("perhaps at about the Pawnee village laid down in the map"). Then he was to go down the river to the Mississippi, cross through Illinois, Kentucky, Tennessee, and Virginia, and return immediately to Philadelphia.

During his travels Nuttall was to look for, make notes about, collect, and preserve a great variety of natural wonders. Barton was most specific in these instructions. For example, Nuttall was to go down "the Ohio to the mouth of Big Beaver, where you will find Frasera, of which I wish a few pounds of good roots . . . procede to Sandusky, where I wish you to stay at least so long as to acquire five good specimens of the Wyandot-language. Here you also may find some curious plants; and don't omit to make inquiries about the disease of goiter . . . whether it is connected with fatuity, etc. . . . I wish you to explore [Lake Winnipeg] with great and minute attention for plants, animals, minerals &c. Obtain a list of the fish of these lakes."

When he got back, Nuttall was to turn over all his notes to Barton. ("This journal," read the contract, "and all the observations you may make are to be my exclusive property, and no parts of them are to be communicated without my consent, to any person.") Nuttall was entitled to "a part of all specimens of animals, vegetables, minerals, Indian curiosities, etc., but these you are not to dispose of without my consent, as they might otherwise fall into hands of persons who would use them to my disadvantage." In return for all of this the good doctor advanced Nuttall about one hundred dollars in expense money and agreed to pay him a salary of eight dollars a month, but *only* if and when he came home and the expedition had been completed satisfactorily.

Beyond reducing Nuttall to the status of an indentured scientific servant, the itinerary proposed by Barton was incredible. Sitting in his study, the stay-at-home savant outlined on the sketchiest sort of maps a trip across two-thirds of a largely wilderness continent which would have taxed the abilities of a Lewis, Clark, Boone, or Bridger. For Nuttall, without frontier skills, equipment, or logistic support, it was preposterous. That he accomplished so much, with little more than single-mindedness as a resource, was to make his one of the great travel accomplishments of his time.

Ignorance does not entirely explain or excuse Barton's part in the affair. A good bit of almost criminal cynicism was involved. Writing about it to an acquaintance, Barton commented that Nuttall appeared to be a delightfully "innocent" young man who, "I have no doubt, should his life be spared, will add much to our knowledge of geography and natural history." The image conjured up is clear and Dickensian, of the Philadelphia sophisticate chortling at the shrewd bargain he had made. If Nuttall did return, he, Barton, would have a sort of monopoly on western natural history data. If the unimportant, unconnected young Englishman was not "spared," Barton would be disappointed but would suffer no loss beyond the few dollars advanced toward expenses.

As things turned out, the Barton-Nuttall arrangement was a classic and heartening example of innocence triumphant over deviousness. Oblivious it seems to everything but the opportunity to naturalize to his heart's content, Nuttall signed the strange agreement and went west to endure the hardship and dangers that followed. Because he did, Nuttall, who cared almost nothing for conventional honors, was to become the most honored naturalist of the time. Barton, for whom recognition was everything, got none and is now remembered as a minor and unpleasant associate of Nuttall's.

Nuttall started off on April 12, 1810, by stagecoach.

"Left Philadelphia, 5 oClock in the morning," reads the first journal entry. "The aspect of the vegetable kingdom still appears barren,

antly yields to the smiling aspect of spring, the trees
'on of a few Amentaceae, & the florid buds of the Acer
...ocked in dormant sleep, and the Herbaceae are almost
universaly embosomed in the earth, with but trifling exception of humble Draba & the foetid Pothos but the apathy of the vegetable is more than made up by the pleasing exertions of the feathered songsters, particularly the plaintive note of the Caerulean Warbler and the chearful lark. Besides these birds I saw the Turkey Buzzard, Wild Pidgeon, Thrasher, Blackbird & the twittering Swallow."

Nuttall went on to record some geological observations in an entry that was very typical of the thousands of others he was to make in his subsequent notebooks. Always there are plenty of details about the animals, vegetables, and minerals he encountered but virtually none about himself or the logistics of the day. Few other adventurers, green or seasoned, could have resisted a few soaring passages about how they felt, what mood they were in as they set out on such a bold and risky trip. Nuttall did.

About the only personal matter he apparently thought worthy of mention and therefore one of the few things we know about what he was like as he rattled across Pennsylvania 175 years ago is that he carried with him plasmodium, the malaria protozoan. He had first been infested by these parasites while collecting in Delaware swamps, and during most of his years in the field he was tormented by malarial attacks. Sometimes he would make brief mention of them, usually to explain why he had not done as much as he planned, or spent extra time in some unpromising place. On this first trip, just on the other side of the Susquehanna, the fever struck him. He noted it by way of introduction to a description of a curious chalybeate spring he found outside Bedford, high up in the Appalachians. After some observations on the geological situation and some speculation about the composition of the water, Nuttall wrote clinically, "I felt very much debilitated by the ague & tho't a draught of these waters might be beneficial. I accordingly drank pretty freely." The next day he continued his observations in this matter. "I had no fit of ague as I had expected, & feel my health recruited greatly, & I think the chalybeate water did me essential service tho' I had at first but a mean opinion of its medical effect."

In late April, Nuttall left Pittsburgh and the stage roads to walk to Lake Erie, 140 miles to the north. It was his first real taste of the wilderness, the path he chose lying along the western eaves of the Alleghenies, through heavy virgin forests and extensive swamps. How he coped—how he found his way, made camp, and cooked—was as usual a detail not worthy of comment. Based on his later performances, the chances are good that he coped badly, spent a lot of wet, hungry nights, perhaps huddled in laurel thickets. Even if not, it was a very hard trip. In the dark, damp woods, malaria came on him again. ("I labored almost without interval under the severest attack of intermittent fever. . . . On Sunday the usual nausea coming on . . . I tried bleeding twice, ineffectively. I tried an emetic, but it did not operate. I then tried Calomel & Jalap.") Also, because of confused arrangements with a teamster, he lost his trunk and had to make a sixty-mile detour to retrieve it, stumbling through swamps and thickets with the box on his head. Among other things, the trunk held powder, shot, five reams of paper for pressing plants, a thermometer, steel pen, a balance scale, and five notebooks.

All of these misadventures are reported more or less as asides. The bulk of his journal has to do with finding, pursuing, and being enchanted by porcupines, hummingbirds, shrews, salamanders, hawks, upwelling springs of crude petroleum, and of course dozens of new plants. There was, for example, a terrible, by any conventional standards, day when he was trudging along "in shady fir swamps," racked with malaria, toting his trunk on his head. Obviously he was a sick and uncomfortable man, but he was an excited and in his way happy one, for he came upon at least four new species of plants which he was later to describe as the running strawberry bush, toothwort, white dogtooth violet, and the blue-eyed Mary, which particularly delighted him.

Nuttall was to have many such days during his next thirty years of wandering about the backcountry. So far as the nuts and bolts of pathfinding and logistics went, he was a wretched, bumbling traveler. However, before he was finished he was to travel as far and hard as any of the more or less professional hero-explorers of the North American wilderness. While doing so he probably saw more of the new lands, in a close, clinical, objective way, than anyone else. He got about and survived, as he did in those first feverish weeks in the Allegheny swamps, by being able and willing to endure much more than most and by being less distracted than almost anyone by what he had to endure.

After six weeks or so of such agonies and ecstasies, Nuttall reached Detroit. There, along with botanizing, he made a few observations about Indians in the area and, befitting a Yorkshireman, a sour, chauvinistic one about the citizens of that frontier post. "Detroit contains about fifteen hundred inhabitants, mostly French people & *Catholicks*, in full possession of all the superstitions peculiar to that religion." From there Nuttall hitched a ride up Lake Huron with a territorial surveyor who was traveling by canoe to Michilimackinac. In that brawling depot of the northwestern fur trade, he apparently met Wilson Hunt, who was gathering men and supplies for the great Astor-sponsored push to the Pacific which would be made the next season. Nuttall attached himself to the Astorians and in comparative comfort traveled south with them through Michigan, Wisconsin, and Illinois to St. Louis. (Thereafter Nuttall paid slight attention to the ridiculous itinerary laid out by Benjamin Barton. By then he presumably had found out it was impossible for a solitary man to manage and also was discovering places and things which interested him more.)

Nuttall overwintered with the Astorians in St. Louis, perhaps picked up a few dollars working in a local print shop; he met Bradbury and did some local naturalizing with him. Then in the spring they started up the Missouri. After the encounter in which Nuttall had stood ready to fire a musketload of seeds at Teton Sioux, the Astorians turned westward toward the Pacific and their bloody fate. Bradbury, Henry Brakenridge, and Nuttall continued on up the river with a smaller fur party to a post in the Mandan country in what is now western North Dakota. The first two men remained there only briefly and were back in St. Louis in August. Nuttall stayed on, working to the west on his collecting trips, enraptured to be the first professional naturalist to see this High Plains country. (Once he was found, half-starved and lost but full of specimens, one hundred miles away from the post, almost into the Yellowstone country.) When fall and the first frosts came to the high country, he caught a ride downstream on one of the last flatboats bringing furs out of the mountains. He was in St. Louis by the end of October but stayed only long enough to pack his specimens. Then he boarded a riverboat to New Orleans. Along the sparsely settled Mississippi there was much to catch and occupy the attention of a naturalist. Twenty years later, writing in his reference work of ornithology, Nuttall was to recall one of the things he saw as he came down the great river out of the Wild West. There is a mythic quality to the passage, a peculiarly moving evocation of times, places, and circumstances from which we now are separated by so much more than a century and three-quarters.

"In the month of December 1811, while leisurely descending on the bosom of the Mississippi, in one of the trading boats of that period, I had an opportunity of witnessing one of these vast migrations of Whooping Cranes, assembled by many thousands from all the marshes and impassable swamps of the North and West. The whole continent seemed as if giving up its quota of the species to swell the mighty host. . . . The clangor of these numerous legions, passing along, high in the air, seemed almost deafening; the confused cry of the vast army continued, with the lengthening procession, and as the vocal call continued nearly throughout the whole night, without intermission, some idea may be formed of the immensity of the numbers now assembled on their annual journey to regions of the South."

In New Orleans, Nuttall, as per his contract, shipped his notes and that share of the collections which he thought proper to Benjamin Barton, but because the War of 1812 was imminent, he himself got one of the last ships leaving New Orleans for his native England. (Barton reacted irascibly to what he regarded as this treacherous change in plans. Very likely he was most disturbed by the fact that Nuttall would not be available to do the desk work of whipping the western material into shape for Barton to publish subsequently. Barton himself never got around to the job.)

Nuttall stayed in England for nearly three years attending to family business, doing some local collecting, and sharing some of his collections of western flora with botanically minded individuals and institutions. One consequence was that in 1813 he was elected a member of the Linnean Society, then perhaps the world's leading floral organization. The honor emphasized an elemental fact about Nuttall which is, at least at this distance in time, sometimes easy to overlook. Despite the dramatic nature of his solitary travels in the North American wilderness, he was not some sort of untutored Johnny Appleseed, a simple "plant lover" mooning about the countryside. By all standards of the time, he was a formal and professional student. He was largely self-taught, but so were most of his colleagues, since the natural sciences were only beginning to be included in academic curricula. However he acquired it, he had a good taxonomic background when he first arrived in the United States in 1808—dilettantes on landing in a strange land do not rush off to find a greenbrier bush and speculate about its relation to the passionflower vine. His subsequent collecting trips had sharpened a skill which was to be remarked on admiringly by many of his peers—an extraordinary, intuitive feel for relationships between plants encountered in the field, far from guide and reference books. For Nuttall, the period he spent in England after leaving the West amounted to something of a graduate course, and by the time he left he was familiar with virtually all of the current works and theories, as knowledgeable a professional botanist as there was to be found on either side of the Atlantic.

On his return to Philadelphia in 1815, Nuttall already was gripped by the ambition to prepare a definitive work on the flora of North America. (Barton died in that same year, freeing Nuttall to write about the discoveries he had made.) He stayed only a few months in the city and then set off, alone as usual, on a series of expeditions through the Appalachians and Deep South. These were to occupy him for most of the next two years. Returning in 1817 to Philadelphia, and having been elected to both the American Philosophical Society and Academy of Natural Sciences, he settled down to classifying and describing what he had found in the field. In midsummer of 1818 he published his monumental *Genera of North American Plants with a Catalog of Species through 1817*. It was the first comprehensive botanical reference to be published in the United States. It was to remain one of the most influential for decades to come and still stands as the great pioneering inventory of American flora. Nuttall not only wrote but set the type for most of the six-hundred-page work, which contained detailed descriptions of 834 genera, some ten percent of which (along with hundreds of species) were new ones. It was so much more extensive than and superior to anything that had been done previously that Nuttall was almost immediately established in the very first rank of American natural historians.

The stir which the *Genera* caused did not seem to impress Nuttall greatly, certainly not enough to distract him long from what he regarded as his real work, continuing to add to the great catalog of American fauna and flora. Just three months after publication, he left Philadelphia for what was then called the Arkansas Territory but included parts of Texas and Oklahoma and stretched for an indeterminate distance westward toward the Rockies. It was rough country and largely unexplored as far as its natural history was concerned. Making the first survey was to be Nuttall's most difficult trip, one of the most difficult any American naturalist was ever to make. Beyond the normal but considerable hardships of wilderness travel and survival, there were problems which might be called social, if the word is stretched far enough. The Arkansas country had become something of a place of exile, a last refuge for a variety of failed, frustrated and ferocious types: pirates, cutthroats, gun runners, a ragtag of Indian tribes, debauched and dispossessed by what was sometimes called "advancing European civilization." Before his year-and-a-half trip was over, Nuttall was discommoded and delayed by an assortment of rascals, ordinary drunks, and bullies.

Nuttall himself had not changed much, was still a stunningly inept explorer but an indefatigable observer. During the spring of 1819, in a kind of classic Nuttall mixup, he lost not only all of his supplies but an entire military escort as well. Much of the territory was then under what amounted to martial law, and therefore for both protection and logistical support, Nuttall had gone up the Kiamichi River with a party of soldiers. Reaching the confluence with the Red River, the company camped and recuperated for a few days. Nuttall immediately commenced naturalizing and shortly became lost for a day and night. By the time he found his way back to the camp, the soldiers had packed up and gone, leaving the naturalist with only what he carried in his hands and very much stranded because "he dared not venture alone and unprepared through such a difficult and mountainous wilderness."

However, the circumstances did not unhinge or humiliate Nuttall as they might have a more conventional traveler, and in fact he came to regard the episode as being more fortunate than disastrous. In time he found a backwoods clearing inhabited by a family named Styles. They were presumably astounded to have a British perfesser come straggling out of the woods, but were also obviously impressed with the curious little stranger. They took him in and kept him for three weeks until arrangements could be made for his return to the settlements. Settling in with the family, Nuttall had, according to his journal, a grand time and spent very little of it fretting about the enforced disruption of his itinerary.

"My botanical acquisitions in the prairies proved so interesting as almost to make me forget my situation, cast away as I was . . . unprovided with every means of subsistence. . . . The singular appearance of these vast meadows, now so profusely decorated with flowers, as seen from a distance, can scarcely be described. Several large circumscribed tracts were perfectly gilded with millions of the flowers Rudbeckia amplexicaulis [a coneflower], bordered by other irregular snow-white fields of a new species of Coriandrum [coriander]. . . . Today [June 6, 1819] I went five or six miles to collect specimens of the Centaurea [a knapweed] which, as being the only species of this numerous genus indigenous to America, had excited my curiosity."

Returning to Fort Smith in July, Nuttall almost immediately headed back into the wilds, this time determined to travel west to the Rockies. He went with a veteran trapper–mountain man by the name of Lee, who had the same general geographical objective. On horseback, pushing through dense thickets and stagnant swamps, tortured by insects, Nuttall became delirious, sometimes fainted in the saddle because of recurring attacks of malaria. On the Canadian River, Lee, alarmed by the weakening condition of his companion, suggested they stop, but Nuttall would not hear of it. By September they had reached the Cimarron, and Lee's horse was lost in the quicksand. This was enough for the frontiersman, who said that no matter what the loss to science, they were turning back. He constructed a rough canoe out of a cottonwood tree and in it floated downstream while Nuttall followed along the bank on the remaining horse. The river water was foul to drink. They had little food and lots of bugs. Nuttall's malarial fevers grew worse. Then they were set upon by a decadent band of Osage

Indians. Not strong enough to be mortally dangerous, the Osages, under the direction of a blind chief, made persistent attempts to steal the few possessions the two white men had retained. After the Osages were finally shaken, the bottomland brush became so thick that Nuttall could not keep pace with the canoe. Lee went on ahead, and for the final two days of the terrible trip Nuttall plodded on alone through mud and violent thunderstorms, racked by chills and fevers, unable to get warm or dry because, as he confessed, he could not start a fire. Despite everything, Nuttall remained the naturalist, though close to being a terminal one, still observing enough to spot "a very curious Gaura, an undescribed species of Donia, of Eriogonum, of Achyranthes, Arundo and Gentian." Finally arriving at a trading post, "My feet and legs were so swelled, in consequence of weakness and exposure to extreme heat and cold, that it was necessary to cut off my pantaloons, and at night both my hands and feet were affected by the most violent cramp."

Though at times it was a near thing, Nuttall was again "spared" in Arkansas and got back to Philadelphia, by way of New Orleans, in the early spring of 1820. His collections in the Southwest plus the success of the *Genera* made him, without close rival, the first American botanist. In consequence he was, in 1822, invited to come to Harvard University as a lecturer in natural history and as the curator of that institution's botanical gardens. He accepted and was to remain at Harvard, off and on, for the next decade, the most settled and conventionally comfortable period he was to spend in the United States. At Harvard, Nuttall had the leisure and inclination to organize his notes on birds and eventually published *A Manual of the Ornithology of the United States and Canada* in two volumes. Again it was the first American-published book of its kind and was to be a standard reference for the next several decades.

During the Harvard years Nuttall made regular collecting trips up and down the East Coast from Maine to Florida and also extended his acquaintance with fellow scientists on both sides of the Atlantic. Among those who came to visit him at Harvard was Audubon, who was both admiring of and sometimes amused by the little Englishman. According to Audubon's later account, the two great naturalists decided to take a bit of a busman's holiday and, one August day in 1832, to stroll toward Brookline. There they were "saluted" by the call of an olive-sided flycatcher, which Nuttall had previously described and wanted to show to his new friend. Nuttall thereupon trotted off to a farm to borrow a gun. "He returned," wrote the wondering Audubon, "with a large musket, a cow's horn filled with powder, and a handful of shot nearly as large as peas; but just as I commenced charging this curious piece, I discovered it was flintless." Finally they were able to borrow another gun more suitable for flycatcher hunting. Audubon eventually bagged the bird, which is now known as *Nuttallornis borealis*. The friendship between the two men continued. On his next western trip Nuttall sent Audubon a variety of specimens. Both the friendship and the gift were memoralized when Audubon used two of the specimens, a pair of band-tailed pigeons, in a painting. The birds were fittingly perched on a Pacific dogwood (*Cornus nuttallii*), also named for the botanist.

Easy and congenial enough though he was with his professional colleagues, Nuttall did not seem to fit in well with the general Harvard community and made little effort to conceal his disinterest in it. For example, he modified his apartment (in a house by the botanical gardens) along semi-fortress lines, which precluded casual visiting. He blocked off what had been the stairway and cut a hole in the first-floor ceiling which led into his study. Under the new arrangement this could only be entered by climbing a ladder. He also walled in the passage to the kitchen, leaving a slit through which, when he thought of it, food and dirty dishes could be passed. General entrance to his quarters was made through a window that faced the gardens. All of which secured Nuttall's privacy but understandably contributed to his on-campus reputation for unsociability.

Despite the honor and ease, Nuttall grew increasingly bored at Harvard and with, as he once commented, "the tittle-tattle of Cambridge." Openly he fretted that he was "vegetating" there. In 1833 a gaudy opportunity to get out of academia and back into the field presented itself. In Cambridge there was a young businessman by the name of Nathaniel Wyeth who had a lively amateur interest in natural history and occasionally visited Nuttall to accompany the master on bird and botanizing walks. Wyeth was moderately affluent, having done well in the ice business, but as restless in the staid Boston commercial community as Nuttall was in the academic one. Therefore, Wyeth had hit on the idea of getting into the western trade, shipping goods in and taking resources (principally furs) out of the Rocky Mountain area by way of a post he would establish in Oregon. Wyeth went west during 1832–33 to scout out prospects, and while there he made a contract with the Rocky Mountain Fur Company to bring a load of goods across the prairie in the summer of 1834 and sell them to the mountain men assembled at their rendezvous along the Green River in Wyoming. Excited by the prospects of both commercial success and adventure, Wyeth returned to the Boston area late in the fall of 1833 and invited his old mentor Nuttall to go with the next summer's party, which would travel from St. Louis to Oregon. Nuttall was immediately and enthusiastically agreeable. Such a trek would permit him at last to see the Rockies, a range he had tried to get to from the Mandan country in the fall of 1811 and again on the trip through Arkansas in 1819 which had almost cost him his life. Though his Harvard salary had been increased, he left that institution without any hesitation and with little apparent regret. He went back to Philadelphia to discuss the project with scientific colleagues there, and in the course of things offered to take a much younger naturalist, twenty-four-year-old John Kirk Townsend, who was principally an ornithologist, with him on the expedition. Townsend was delighted; the two men traveled to St. Louis that winter and in April 1834 set off toward Oregon with the Wyeth party.

Seeing the country for the first time and being a more conventionally demonstrative journalist, Townsend wrote ecstatically in his diary as the group started up the Missouri:

"On the 28th of April at 10 o'clock in the morning, our caravan, consisting of seventy men, and two hundred and fifty horses, began its march; Captain Wyeth and Milton Sublette (veteran fur trader) took the lead, Mr. N. and myself rode beside them; then the men in double file, each leading with a line two horses heavily laden. . . . It was altogether so exciting that I could scarcely contain myself."

Nuttall of course had no such emotional comment to make, and in fact by then had more or less stopped keeping any sort of a narrative journal, confining himself to strictly technical notes which he needed to describe the specimens he found. Though he had collected along the same riverbanks nearly a quarter of a century before and others had come after him, he had no difficulty adding to the great inventory of plants. "Mr. N.," wrote Townsend of their first days along the river, "is finding dozens of new species daily."

The two naturalists also did well with the birds. Very shortly, Townsend found the previously undescribed chestnut-collared longspur, and Nuttall looked up from his plants long enough to add another new bird, the mourning sparrow. (Ten years later Audubon was to refind and rename the species, calling it the Harris' sparrow.) By the end of May the caravan had reached the vicinity of Scotts Bluff, near what is now the Nebraska-Wyoming boundary. There they passed through a deep, narrow coulee carpeted with spring flowers. Townsend wrote:

"It was a most enchanting sight; even the men noticed it, and more than one of our matter-of-fact people exclaimed, beautiful, beautiful! Mr. N. was here in his glory. He rode on ahead of the company and cleared the passages with a trembling and eager hand, looking anxiously back at the approaching party, as though he feared it would come ere he had finished and tread his lovely prizes underfoot."

Wyeth's immediate objective was the trappers' rendezvous on the Green River where he planned, per the contract of the previous year, to sell the goods he had packed over the prairies to the assembled mountain men. Arriving in late June, he found that the hard-bitten partners of the Rocky Mountain Fur Company had brought in their own trade goods, and they told Wyeth in no uncertain terms what he could do with his supposed contract and his Boston business ethics as well. While this was going on, the party remained at the rendezvous site, and the carryings-on were an eye-opener for the two gentle naturalists. The rendezvous system had been organized in the 1820s as a means of getting the trappers' furs at rock-bottom prices. To take some of the sting out of the transaction, the St. Louis hustlers laid on a bash for the mountain men. By 1834 it had grown into a mighty, month-long American saturnalia at which the trappers gambled recklessly, drank incessantly, bragged outrageously, fought, wrestled, gouged eyes, set rabid wolves on each other, generally caroused until they were absolutely dissipated and broke; then, crapulent almost beyond belief, headed back to the lonely mountains to go at the beaver for another eleven months. Wyeth's party camped at the edge of the rendezvous grounds but nevertheless saw enough of the fun and games.

"This being a memorable day," wrote Townsend on the fourth of July, "the liquor kegs were opened. We, therefore, soon had a renewal of the coarse and brutal scenes of the rendezvous. Some of the bacchanals called for a volley in honor of the day. We who were not 'happy' had to lie flat on the ground to avoid the bullets which were careening through the camp."

Thoroughly hornswoggled by the mountain men, Wyeth decided to press on toward Oregon, hoping to dispose of some of his trade goods either on the way or when he arrived on the coast. They followed parts of a route which ten years later was to become famous as the Oregon Trail. It was inevitably a hard and dangerous trip, but the practical difficulties which engaged the rest of the party seemed to be of precious little concern to the two naturalists. Trying to make a little trade and also to protect themselves from the formidable parties of Blackfoot Indians, Wyeth and his men stopped on the upper Snake to build a small post, which they called Fort Hall. While they were doing so, Townsend, among other wonders, bagged a theretofore unknown mountain plover. Nuttall triumphantly found a new nightjar, the poorwill (*Phalaenoptilus nuttallii*), and observed, as though he were in a quiet Harvard garden, the western marsh wren, "a remarkably active and quaint little species, skipping and diving about with great activity after its insect food."

From Fort Hall, short on food, with the horses dying, the party straggled across the awful lava fields above the Snake River, plains of murderously sharp rocks, almost devoid of water and grass. Nuttall found this all quite interesting, was very curious about "the beautiful pebbles of chalcedony and fine agate." However, there is circumstantial evidence that even he was beginning to be affected by the general hardships. In regard to a species of hawthorn, Nuttall noted, "the sweet berries were also welcome food." On September 2nd, a day away from the Columbia River and the relief it promised, Townsend shot a small owl, but it was not preserved for science because "Mr. N." roasted and ate the bird.

Finally reaching the mouth of the Columbia River, Nuttall, while the rest of the party recuperated and made provisions for shelter and trade, rolled up his pants and waded out into the ocean. There he scooped around and came up with some new shellfish, among them a creature which thereafter was to be known as Nuttall's cockle. Though there is no indication that he planned or thought of it that way, it was a symbolic act and find, marking as it did the first passage of the entire continent by a working naturalist.

Between them, Nuttall and Townsend had collected and dragged across the Rockies thousands of specimens, animal, vegetable, and mineral. Concerned about the effects of a damp Oregon winter on these treasures, Nuttall, always inventive when it came to protecting his natural history interests, found them passage on a schooner bound for Hawaii. The two men stayed the winter and of course did more collecting—Nuttall becoming particularly interested in the ferns of the Hawaiian Islands. The next spring he came back to bird and botanize along the coasts of what are now Oregon, Washington, and British Columbia, but once more returned to Hawaii for the winter. In the spring of 1836 Nuttall finally started home, sailing first to San Diego to wait for the Boston-bound brig *Alert*. Having time on his hands, he began wading about the harbor and eventually was able to collect twenty-one new species of shellfish and fifteen new crustaceans. When the *Alert* finally came into port it had among its crew a wandering young gentleman-adventurer, Richard Henry Dana, who had been a student at Harvard while Nuttall was a professor. Later, when Dana wrote his own classic traveler's tale, *Two Years before the Mast,* he was to describe this improbable meeting:

"I had left him [Nuttall] quietly seated in the chair of Botany and Ornithology in Harvard University, and the next I saw of him he was strolling about San Diego beach, in a sailor's pea-jacket, with a wide straw hat, and barefooted, with his trousers rolled up to his knees, picking up stones and shells. . . . The crew called Mr. Nuttall 'Old Curious' from his zeal for curiosities, and some of them said that he was crazy, and that his friends let him go about and amuse himself in this way. Why else a rich man (sailors call every man rich who does not work with his hands) should leave a Christian country, and come to such a place as California, to pick up shells and stones, they could not understand. One of them, however, who had seen something more of the world ashore, set all to rights, as he thought: 'Oh, 'vast there! You don't know anything about them craft. I've seen them colleges and know the ropes. They keep all such things for cur'osities and study 'em, and have men a purpose to go and get 'em. This old chap knows what he's about. He a'n't the child you take him for. He'll carry all these things to the college, and if they are better than any they have had before, he'll be head of the college. The old covey knows the ropes. He has worked a traverse over 'em and come 'way out here where nobody's ever been afore, and where they'll never think of coming.' "

This portrait, substantially so similar to the one contributed by Henry Brackenridge of the young Nuttall in 1811, was the last of "Old Curious" in the field. Thereafter his pursuits were to be more gentlemanly and comfortable ones conducted in tamish places. Having finally naturalized his way to the western ocean, whatever drove Nuttall seemed to have been soothed and satisfied. Also, of course, his final, nearly three-year-long field expedition more or less ended the great pioneering period for everyone. Those who followed were to find much that was new and important, but essentially they filled in a naturalist's map of the country which Nuttall had outlined.

Returning to Boston, Nuttall was widely acclaimed. (For example, the master of the *Alert* who brought the naturalist home refused to accept passage money from him, saying that he had not traveled "for his own amusement but for the benefit of mankind.") Thereafter, based principally in Philadelphia, Nuttall spent several years organizing his western collections, publishing papers, working on technical problems of taxonomy. He seemed to become a more social man, at least in comparison with his former habits, traveling up and down the seaboard, visiting and being visited by colleagues. He even attended at least one wedding—in the company of Henry Wadsworth Longfellow—at which, as far as is known, neither the bride nor groom was a naturalist.

Toward the end of the decade he began to dispose of his collections and generally to put his affairs in order, preparatory to leaving the United States permanently. In England his uncle Jonas, retired from but affluent because of the printing business, had died, and Nuttall had inherited his uncle's large estate, Nutgrove, a few miles outside Liverpool, as well as the means to maintain it and himself as a country squire. There was, however, one restriction, reflecting the uncle's jaundiced view of his nephew's profession and proclivities. To keep the legacy it was necessary for Nuttall to live for at least nine months out of every year in England.

The last major work of his professional career was completed just before he left the United States, a three-volume "appendix" updating *The North American Sylva,* a work originally authored by François André Michaux at the beginning of the century. In a preface to it, Nuttall made a formal and fond (also, for him, an uncharacteristically emotional and biographical) farewell to North America:

"Thirty-four years ago, I left England to explore the natural history of the United States . . . after a boisterous and dangerous passage, our dismasted vessel entered the Capes of the Delaware in the month of April. . . . All was new; and life, like the season, was then full of hope and enthusiasm. The forests, apparently unbroken in their primeval solitude and repose, spread themselves on either hand as we passed placidly along. . . . Scenes like these have little attraction for ordinary life. But to the naturalist it is far otherwise; privations to him are cheaply purchased if he may but roam over the wild domain of primeval nature. . . . For thousands of miles my chief converse has been in the wilderness with the spontaneous productions of nature: and the study of these objects and their contemplation has been to me a source of constant delight. This fervid curiosity led me to the banks of the Ohio, through the dark forests and brakes of the Mississippi, to the distant lakes of the northern frontier; through the wilds of Florida; far up the Red River and Missouri, and through the territory of Arkansas. And now across the arid plains of the Far West, beyond the steppes of the Rocky Mountains, down the Oregon to the extended shores of the Pacific, across the distant ocean to that famous group, the Sandwich Islands. . . . But the 'oft-told tale' approaches to its close, and I must now bid a long adieu to the 'New World,' its sylvan scenes, its mountains, wilds and plains; and henceforth, in the evening of my career, I return, almost an exile, to the land of my nativity."

As he seemed to foresee, his serious professional work was finished when, at fifty-five, Nuttall left Philadelphia late in 1841. His life, however, was not, as he was to live eighteen more years as an English country gentleman. He remained an interested onlooker as far as the natural sciences were concerned, but never again a pioneer, a mover and shaker. He occupied himself with managing the estate and working as a hobbyist in its extensive gardens and greenhouse.

Sometimes in considering the life of Nuttall there is a tendency to pass lightly and quickly over the last decades as if they somehow detract from the previous ones, suggest he spent the vigorous years in America marking time, awaiting a legacy. Yet viewed in another way, this long, gentle autumnal period seems to be another reflection of a truly rare personality, of a man of uncommon mind and character.

Had Nuttall wanted wealth and position, he could have had them much earlier and more easily by listening to his uncle and staying home to mind the family business. Instead he gave himself over to an improbable passion, and despite hardships and deprivations which most would have found literally unbearable, he pursued it across the whole of a wild continent. Then there came a time—upon his return from California—when it seemed he had decided that the quest was finished for him, that he had done what he had so single-mindedly set out to do. Rather than hanging on querulously as his powers and zest diminished, jealously contending with new men and ideas, he stepped aside, secure in the sense that his accomplishments—what he had seen, classified, and named—were unique and would forever remain unique.

All of which is the sheerest speculation, such matters being of the sort Nuttall would and did keep private. However, simply as a matter of historical record, Thomas Nuttall was obviously a man driven by a great passion but also a man who, for whatever reasons, was able to retire from it with grace and dignity. This may have been the final and one of the greatest eccentricities, for such a course of action is rare.

At Nutgrove, Nuttall was able—as he was not when driven by his passion—to develop family ties and feelings. A particular favorite was a nephew, Jonas Booth, an adventurous young man with a considerable enthusiasm for botany. With the aid of his famous uncle, who provided necessary introductions, Booth arranged to visit the Himalayas with a British collection expedition. There he found among other things a variety of Asiatic rhododendrons, from which he sent seeds and cuttings back to Nutgrove. Propagating these exotics from forests he would never see became one of Nuttall's principal interests and pleasures. He sent one of the most spectacular of the bushes to Kew Gardens. In the summer of 1859 this specimen flowered for the first time. On the sixth of September, Nuttall wrote to the curator expressing his special satisfaction about this botanical event.

The plant burst into bloom at a particularly appropriate time. Four days later the old naturalist, for whom it was named *Rhododendron nuttallii,* died.

Trophy Foraging

I grew up in the vicinity of Kalamazoo, Michigan, but had relatives scattered all over that state from Buchanan to Bad Axe to Ishpeming. Among the most impressive was an elderly great-uncle-in-law by the name of Deb, who long before I knew him had more or less abdicated conventional family responsibilities and lived most of the time in a cabin on an island in the St. Joseph River. He occupied himself hunting, trapping, fishing, making a little liquor and poaching (in those days there were in Chicago—only 100 miles away—certain restaurants and gourmets who fancied wild delicacies enough not to ask embarrassing questions about seasons and sources). Also, as I learned by eavesdropping on the horrified gossip of my elders, when Deb got a bit ahead in the hard-cash department he was inclined to roar into the settlements to play a little cards or horses or worse.

Naturally, the solid citizens of the clan thought Deb a scandal, their own shameful version of Pap Finn. I admired him greatly, however, and, whenever I could escape from polite society, spent as much time as possible on the island. On and around the river I learned a lot of interesting skills and habits: to trap, butcher, fry and enjoy eating snapping turtles; to set trotlines; to ferry a canoe across the current; to clip the wings of a mallard duck; to imitate a wounded rabbit by kissing the back of the hand, a sound that attracts hawks, owls and foxes; to use a .22 rifle, the poacher's friend; to chew tobacco; to play cribbage and pitch.

Another thing we did (and something which has proved to have considerable carry-over value for me) was to forage—that is, rummage about in the bush looking for wild edibles. The motive for foraging is similar to that for hunting but involves raiding the flora rather than the fauna. In part because of Deb's good example and instruction, I developed early a taste for a variety of feral fruits and vegetables, fungi, roots, stalks, stems and seeds, but perhaps even more important, a taste for the act of finding them. For 40 subsequent years I have been indulging both these appetites.

Recently there has been a considerable revival of interest in the ancient pastime of foraging, probably because it fits in so nicely with a lot of other contemporary conceits and concerns—the Green Sensibility, Organic (as opposed to old-fashioned inorganic) Feeding, the New Hypochondria. Whatever the reasons, there seem to be a lot more people than ever before wandering about the countryside nibbling on barks and weeds. Concurrently there has been a great outburst of highly serious rhetoric about the meaning and mystique of eating wild. Gurus of the new foraging chic will insist that, say, yellow pond-lily roots will keep down hangnails, spruce up the karma and induce gastronomic orgasms. Some of these claims are about as hard to swallow as a green persimmon. For starters, expertly prepared pond-lily roots taste somewhat like weak, watery, muddy potatoes laced with Styrofoam. Still, there is certainly no harm in people thinking they will achieve union with the ecological absolute by eating dandelions. However, such notions create a lot of heavy metaphysical luggage that is not needed for such a simple trip.

It has always seemed to me that what it all boils—and sometimes bakes and stews—down to is that you forage for about the same reasons you run around trying to collect the semifinalist trophy in the Class C Over 45 doubles tournament of the Chop and Lob Racket Club. It makes little sense, but it is fun. Also it is moderately stimulating exercise. (According to the latest report of the International Institute for Applied Foraging Research of Iron Springs, Pa., five sets of August-afternoon tennis have the same strenuosity factor as preparing two quarts of wild blackberry jam from brier bush to Mason jar.) It is a good way to escape for a few hours from whatever needs escaping. The tangible rewards are often purely symbolic—yellow pond-lily roots being no better as food than $6.75 tennis cups are as decorative sculpture. However, occasionally there is a foraging bonus. The trophy you bring back may have considerable intrinsic value—not monetary but sensual, as in the case of, say, the oyster mushroom, of which more in a moment. In any event, these are reasons why foraging has always seemed to me to be as worthy as any other kind of recreation and no more absurd.

All of which prefaces the hard news that last year was one of the finest foraging years on record, at least in the central Pennsylvania highlands where I have been located for two decades. As a matter of fact, exceptional foraging was just one, though perhaps the most satisfying, of a number of unusual phenomena. For example, the winter was the most memorable in Pennsylvania since Valley Forge. It ended abruptly in early May shortly after the last snow when a searing August-type drought withered spring corn and pastures. The drought was broken by a rainstorm of a volume that the U.S. Geological Survey says should occur only once every 500 years. Then one morning there was an unidentified underground racket that shook local people out of their beds. Experts said an immense limestone cavern had collapsed underneath us, but naturally there has been a lot of scary and private explanation about what really happened on "The Morning the Earth Moved." It snowed again in October but was balmy enough to swim in quarries in November. There was an unverified report of a 42-pound raccoon and, later, monster sightings—one of a Bigfoot type and the other of a lion-style monster.

Some foragers speculate that there may be a natural connection between the earth moving (or the conditions that made it move) and the most abundant crop of fat blackberries anyone can remember, but all that is really known is that a lot of wild desirables did exceedingly well. In consequence, so did a lot of us who enjoy looking for them.

A stray late persimmon aside, my own foraging season generally begins when the sap starts rising in the dozen or so maple trees that grow along a spring drain that meanders down the slope of an abandoned pasture adjacent to my house. Hereabouts this normally occurs anytime between late January and mid-March, but given the ferocity of the winter, there was some concern about if—as well as when—the maples would thaw out. However, there was no need to worry. In mid-February there came a string of clear days when the temperature rose above the 50s at midday and fell to below freezing at night. There is no exact mechanical explanation, but everyone who has had anything to do with maples knows that this kind of brisk alternation between warmish days and cold nights acts like a tonic on the sugar trees and sets their juices to flowing at a great rate. Like the migration of birds, the flow of sap is something we have been collectively watching for a long time without ever precisely understanding the principles involved.

From a safe, contemplative, abstract distance, a maple in the spring comes on as a kind of marvelous arboreal Shmoo. Here stands a tree that nobody has had to buy, plant, weed, prune or spray. Entirely of its own free will it begins to course with a rare and sweet fluid that makes the best pancake topping known to man and costs $11 a half gallon when bought at Uncle Ezra's Vermont Maple Market. All of which from time to time has set people to thinking that the sugar maple is living proof that Barry Commoner was wrong, that indeed there is such a thing as the free lunch. This is the great maple shuck; in fact, it is the great foraging illusion. If the matter—in this case maple syrup—is pursued, reality shortly becomes apparent. It is that Commoner was right—there is no such thing as the free lunch.

There are some peripheral but genuine considerations. If you are going to leech a tree, you have to pierce a considerable trunk, then insert taps, which have to be made, borrowed or bought. The end of these taps must be fitted with some kind of container suitable for catching and holding whatever leaks out of the tree. All of this leads up to the central reality. What comes out of the tree is not pancake dressing but maple sap, a thin, clear, watery liquid. The goodness is there (it is very faintly sweet) but to get at it you have to boil the sap until a lot of water has disappeared and very little syrup remains. Generally speaking, 35 gallons of sap evaporates down to a gallon of syrup, although the amount varies from place to place, tree to tree, even day to day.

Boiling presupposes heat, which requires fuel. Anything flammable can be used—coal, gas, electricity, even old truck tires (the favorite combustible of another local syrup maker)—but wood is the classic material and maple somehow the most appropriate wood with which to fuel the fire to reduce maple sap. Whatever the fuel, it obviously has to be collected, prepared and lugged to the fire.

Maple sap may be mostly water but making syrup out of it is considerably trickier than boiling water. If you don't boil out enough water, you are left with a thin, semisweet concoction that promptly ferments into a sour, vinegary one. On the other hand, if you turn your back for a few minutes on a boiling sap pan, all the water will vanish, leaving you with a kind of caramelized epoxy, good only for patching canoes. To avoid these undesirable extremes, efficient commercial operators use manufactured evaporator devices equipped with gauges that indicate precisely when the proper sugar content has been reached. However, this is too easy and tame for a genuine recreational forager. Our system is intuitive—it involves peering through flames and smoke to monitor the color of the brew, letting it run off a wooden spoon to test viscosity and finally tasting the boiling sap. In the spring you can dependably identify amateur syrup makers by their red eyes, scorched jackets and blistered tongues.

Personally, I have come up with only one syrup-making refinement that seems worth making public; it is that operations will be easier and more entertaining if you get yourself a women's tennis team. Several years ago I was more or less bequeathed such a sporting group by the athletic director of a small local college. Because both tennis players and maple trees tend to become active at about the same time of year, there was some concern at first that syrup and tennis administration might prove to be incompatible interests. However, they are, in fact, complementary, chiefly because it turns out that modern college women can be convinced, or think it politic to be convinced, that prancing about a maple grove is a splendid preseason conditioning exercise. Generally, singles candidates are best employed gathering sap and bringing it back to the outdoor factory. Lugging five-gallon buckets in each hand builds upper-body strength, while hopping about amongst the rocks and brush piles improves agility and the ability to get down low for ground strokes. Doubles players cannot, of course, be expected to show such mobility, but when paired, they make very good stove stokers and boiling-sap tasters, these being exercises that teach cooperation and how to remain cool under fire.

If simple labor were all that was involved, a men's team might be as useful, but there is a special factor that makes women considerably superior. It is that they tend to wear, and wear out, a lot more panty hose than the average man. Converted panty hose are the best known filters for sap, which has to be strained many times between tree and syrup bottle. Tennis players, at least those who are ambitious of being included on traveling squads, will cheerfully save up their defunct underwear and bring it to the old coach while the sap is flowing and before the varsity team is selected.

More or less in this fashion we produced seven gallons of syrup, a new local foraging record. Figuring no more than minimum wages for amateur athletes, it cost about $25 for each gallon. As to quality, more adventurous feeders may find the home brew interesting. It tends to be black and tangy by reason of the wood smoke it has absorbed. Like a Cracker Jack box, it often contains surprises—bits of charcoal, twigs and an occasional caramelized wasp that has slipped through a rent in a filter. (The modern athletic college woman is terribly hard on panty hose, and wasp-sized holes are not unusual.)

All of which illustrates one of the central principles, or paradoxes, of foraging: you do not make syrup solely, or perhaps not even largely, to get syrup. If this is the only objective, then it is best (i.e., easier, quicker and cheaper) to go on about one's regular business—milking cows, robbing banks, writing magazine articles or whatever. Part of the proceeds can then be taken to East Sawed Log, Vermont, and exchanged with Uncle Ezra for some of his quality-controlled golden syrup.

The truth is that syruping and a lot of other foraging activities balance out only when the "other rewards" column is totaled. If you are so inclined and don't have to do it as real work, then it can be invigorating to spend a December day cutting and stacking wood for subsequent syrup fires. Slogging around on snowshoes setting taps brightens up a January weekend. Waiting for the sap to flow gives a nice anticipatory zest to the next few weeks. It is a pleasure to go around talking not necessarily to, but at, maple trees, praising the very sappy ones, admonishing the laggards. Sitting around an open fire on a cold March night watching boiling sap and tennis players is warming in many ways. Just why is essentially unfathomable, in much the same way that it is not easy to explain why it feels good to serve four aces in a row. If someone comes up with a good explanation of the former, it will probably do as well as an explanation of the latter.

About the time the taps have been pulled from the maples and the syrup pans either scoured or junked because they are beyond scouring, it is time to start after mushrooms. This is a major event on the foraging calendar and a very serious one, the wild mushroomer being under a lot more pressure to succeed than, say, the syrup maker, because if he fails in his quest he cannot fall back and feed on more or less comparable storebought produce. Such an assertion may raise some skeptical questions—such as, what are those round, rubbery things, dished out in steak houses, on pizzas and in omelets, that are advertised as mushrooms? They are commercially raised mushrooms and they compare to many wild species as orangeade does to freshly squeezed orange juice. The trouble is that though a lot of attempts have been made, nobody has found out how to propagate dependably and thus to peddle profitably the tastier feral species. Commercial growers, therefore, concentrate on a variety of meadow mushroom called *Agaricus,* which has permitted itself to be domesticated. *Agaricus* is a good enough food, but I have yet to meet anyone who has tried both wild and tame mushrooms who does not prefer the former by a wide margin. Thus there are a lot of strongly motivated mushroom hunters.

Among the 3,000 or so species of fungi in this country, an undetermined number are, as they say in the guides, "edible and choice." Unfortunately, another group can give you headaches, stomach cramps, hallucinations or make you dead. Deciding which is which can be tricky and enlivens this pastime. As a practical matter, most foragers, rather than puzzle over every species, settle for a few of indisputably benign properties. On my own good list there are about 20 mushrooms that I get and gobble whenever met, but of these, three are superior. They are morels (*Morchella esculenta*), chanterelles (*Cantharellus cibarius*) and oysters (*Pleurotus ostreatus*).

Unfortunately, in an otherwise splendid foraging year, about the only plants that didn't do very well were the morels, which normally are among the first good mushrooms to appear in the spring. Last spring they simply weren't there, a failure noted generally by everyone. Normally secretive and competitive, mushroomers spent a lot of time poor-mouthing among themselves, theorizing about what had happened. Some thought the hard winter had done in the morels, while others held the unseasonably hot spring responsible, but nobody really knew. Maybe the monsters got them.

Whatever the reason, blessedly it affected only the morels. Chanterelles, yellow vase-shaped fungi that grow best on mountainsides under heavy deciduous or evergreen shade, were in at least regular supply. Oysters were in absolutely marvelous abundance. Choosing among these three species solely on gastronomical grounds is silly, like trying to rate ambrosia, nectar and manna. But for a variety of practical foraging reasons I am inclined, especially after last year, to give oysters the nod. They are cosmopolitan, being distributed across much of the country, and have a long fruiting season, appearing in the spring, slacking off in the hot weather, but coming on again strong in the fall and lasting until early winter. Last year had a lot of cool, moist weather, which they like, and oysters were there for the foraging in all but three months—January, February and July.

Oysters got their common name because they are usually dead- to cream-white in color, ovalish in shape, fleshy in substance. Individual caps are large—up to a foot across—and they grow, mostly on dead wood, in clusters of a dozen or more caps. Consequently, they are very easy to spot, and once you have found a colony it will always provide at least one meal and often several. You can do with oysters what you do with any mushrooms—put them in soups, salads or hors d'oeuvres, but we purists deride such mincing uses. All that is necessary is to fry them in butter, for 10 minutes or so, until the moisture has stewed out and the edges begin to brown and curl. So prepared, three or four of the big caps will make a substantial main dish that tastes something like a combination of good steak and fresh shellfish, and provides all the explanation any rational person needs as to why there are such crowds of wild-mushroom hunters.

The mushroom is a classic example of a wild thing that is not only edible but also preeminently worth eating. There is a lot of other flora that, while technically possible to ingest without harmful effect, gives very little sustenance or pleasure. As far as I am concerned, there is a whole mess of late spring and early summer herbage that falls into this category: knotweeds, pigweeds, lambs'-quarters, milkweeds, land cresses, plantains and a whole tedious assortment of docks. A good many contemporary authorities are very big on these weeds and have touted them in books, on TV and, for all I know, in art movies—perhaps because there are so many of them that a few remarks about each will give nice heft to a manuscript. Artistic value aside, most of this greenery tastes like either bitter lettuce or bland okra. Also, it is so common as to provide little in the way of exciting foraging.

Before setting forth to browse on such plants it is well to review certain historical facts. People have been into eating for a long time, during which they have investigated much of the flora, captured all they could of it, and put it into gardens and farms. The edible species left in the wild tend to fall into one of two broad categories. First, there are things like mushrooms that are very desirable but have stubbornly resisted domestication. Second, there are plants like, say, curly dock that could easily be domesticated (in fact, they are always forcing their way into gardens as weeds) but have not been because they make such sad, poor food. A good question to ask these herbs is, "Why hasn't the Jolly Green Giant caught you?"

So as not to be dismissed as an inflexible green-vegetable bigot, I will gladly admit there are exceptions. Mixed with cream cheese, watercress makes a passable sandwich, and it grows in boggy places that are usually a pleasure to visit. Day-lily buds, boiled and buttered or fried in batter, are not bad. Both are at least semidomesticated, the former being available in better fresh-vegetable markets and the latter in Oriental specialty shops. A more curious exception is poke, which grows everywhere, including through stone walls. In the spring it sends up big bundles of tender green shoots that are at least as good as asparagus but have a nice peppery taste all their own. Considering how hardy, prolific and tasty it is, it would seem that the Giant would have caught the poke long ago. In parts of Europe this native North American plant is grown in gardens. Here it is generally left in the wild, probably because it scares a lot of people. The difficulty is that while poke shoots are agreeable, they turn ugly as they age. The mature stalks, leaves, berries and, especially, roots contain phytolaccin, a drug with both cathartic and narcotic properties, which, in theory, is potentially lethal.

Before we leave this interesting if formidable plant, another foraging attribute of poke should be noted. Late in the fall, when there is not much greenery, poke roots should be dug up. If nothing else, this is a very good exercise, because the roots are about the size of a Pekingese and have a great talent for wrapping themselves around rocks. The crown of the excavated root will be peppered with a lot of little green knobs. If the root is taken into a warm house or basement, planted and treated more or less like an indoor lily bulb, these bud knots will start sending up shoots. They can be picked (after which more will appear) and eaten, giving welcome relief from the spongy Florida and Texas truck produce that is about all that is otherwise available during northern winters. If the shoots shoot faster than they can be eaten, that is all right, too. They will turn into large, coarse houseplants that are no uglier than philodendron.

Certain feral rabbit foods aside, during the nice summer days, respectable foragers concentrate on berries, for several good reasons. In the first place, though they may not be as superior to their domesticated relatives as wild mushrooms are to tame ones, feral berries are still better. It is well to remember that the object of most domestication is to maximize quantity and appearance, rather than quality. To get bulky, cosmetically appealing berries, the taste nodules are often compressed or eliminated.

The thorns on most wild bushes are much thornier than those on tame plants, but this is not all bad. Brier patches, even in very settled places, often amount to mini-wildernesses. In the middle of a stand of ferocious blackberry canes one is seldom troubled by salespersons, muggers or pollsters. They are also fine places to become better acquainted with nesting birds, woodchucks, snakes and a lot of interesting bugs.

In addition, unlike a lot of foraging that is solitary (either because practitioners cannot get anyone else to share their passion or because they don't *want* anybody along with whom they may have to share what they find), berrying, which goes better the more hands you have, has a long social tradition. After the hard work of planting corn and taking scalps was finished, the Iroquois made a festival out of berry picking, using the occasion for gossiping, cracking jokes, arranging treaties and plotting wars. More recently, but still a long time ago, there were always a couple of fine midsummer days, at least in the St. Joe River valley of southern Michigan, when extended family groups would take an informal holiday, pack a picnic and go off into the brambles to pick berries and have a little fun.

Such customs have deteriorated, but even so, during the lush berry season we were able to put together a congenial party of mixed pickers: two daughters, one daughter's paramour and three golden retrievers. The daughter-paramour team was handicapped, having only two free hands for the unit. The unattached daughter's concentration was poor by reason of her having to pay close attention to a disc jockey who was giving important announcements via portable radio about something called the Top 40. The dogs, however, did very well. Frightened of grouse, unwilling to fetch anything more useful than wet sneakers, they are splendid berry retrievers. If they have to, they will worm their way deep into the briers, nipping off and gobbling down the fattest berries, although they prefer to lurk about until a berry pail is left unguarded, they muzzle in as if it contained sweet kibble.

Any discussion of how to pick berries would seem as superfluous as the sort of thing that now regularly appears in large books with titles like *The Joy of Backpacking* that give technical advice on how to walk. Any child, woman, man or dog who cannot figure out how to get a berry into a bucket or a mouth has a severe problem. Nor is there much need to dwell on what to do with berries once you have them home. They are good plain or with cream, and are even better if pureed (which removes their most objectionable feature—lots of seeds) and poured over ice cream. You can make jams, jellies, wines and cordials out of them, according to formulas that any decent cookbook or great aunt will provide. As a sample, courtesy of a very decent sister-in-law: set aside two quarts of ripe, wild, unwashed raspberries. Boil four cups of sugar in three of water; cool to room temperature. Pour mixture over the berries and mix with a fifth of vodka. Cover the container—cheesecloth will do but panty hose are better. Let it alone until at least Veterans Day and then drink with pleasure, but some care.

Wild berries are low in natural acids. As a result, almost any concoction made with them seems to be improved by adding a few shots of fresh lime juice. Another personal culinary observation is that it is often rewarding to substitute honey for all or part of the sugar called for in recipes.

Among other improbable kin, we have a sister and brother-in-law who some time ago dropped out of the eastern academic world and retreated, or advanced, to the Missouri Ozarks, where they have become upwardly mobile beekeepers. Now we are able to arrange trades—smoky maple syrup for the best southern highland honey a bee ever made. Of course, this is tame, easy-to-come-by honey. The wild, hard way is to go out foraging for it, an urge that overcame several of us one hot August afternoon when everyone was restless for lack of adventure and moments of truth.

Collecting wild honey is a borderline activity, like frog gigging, halfway between hunting and foraging, and perhaps should be excluded from this report. A better reason for not dwelling on subsequent happenings is that they were as humiliating as they were painful. Briefly, we set off to a nearby cabin that once belonged to a moonshiner. One wall of it had been occupied for several seasons by an active swarm of honeybees. When we were done we had a ruined cabin wall, a small brush fire (in hopes of creating a smudge), 27 beestings in assorted human hides, and a faithful but stupid golden retriever with one eye swollen shut. We also had, after sticks, roofing nails and a dead mouse were removed, a gallon of honey. It was molasses-colored and semifermented, but it added a genuinely wild element to berry recipes.

All the local wild berries—straw, dew, black, rasp and wine—did exceptionally well last year, but the elderberries did best of all. In mid-August the tall, spindly bushes were bent double by the weight of heavy clusters of fat, dark purple fruits. There is an odd thing about elderberries—fresh and raw they are awful to nauseating. They have a used taste, like old, musty medicine. Cooking dissipates this unpleasant flavor. Elderberries are prime jelly material, combining excellently with a lot of other fruits, particularly wild grapes and May apples. A good bit of elderberry vinegar is unintentionally made by people attempting to make elderberry wine. Elderberries can be dried in the sun. Properly withered, they look like mouse droppings, but they can be stored indefinitely and used with good results in chutneys, stews and baked goods.

We tried all these things that fine season but the best elderberry concoction was a new one. We should have known about it before, because all the ingredients were familiar, but we didn't and might never have known except for a tip in the published works of the late Euell Gibbons. Now Gibbons was a knowledgeable and indefatigable forager—and a central Pennsylvania one, too. At times he was a bit overenthusiastic, particularly when it came to green weeds, but on elderberries he was, if anything, too subdued.

First you need elderberry juice, which you get like any other fruit juice—mushing a bunch of stewed berries through a filter. Then you get a bushel or so of bright red sumac berry spikes. Scarlet sumac, which grows everywhere in poor, abandoned corners of the land, has a bad name because a much different bush is called poison sumac. However, the scarlet species (*Rhus glabra*) is harmless. If the berry spikes are swished around in a tubful of water for a time, they yield a very drinkable infusion. (After swishing and before drinking, the runoff should be strained. Sumac berries are hairy, and the hairs will clog up around the soft palate if they are not removed.) Straight, it is an interesting drink—astringent, lemony by reason of its malic acid content, but quite thin. It was obviously intended to be mixed—as Gibbons knew—with elderberry juice (about three to one in favor of the sumac), because elderberry juice has a very full flavor but lacks the acid tang. Together with as much honey as seems good, you have what might be called sulderade, and also very nearly the perfect drink. It is the kind of thing that around here is said to whiten the teeth, sweeten the breath and make childbearing a pleasure.

After they have finished with berries, a good many forgers go nutting, but I am not one of them. I have nothing against nuts—in fact, am very much in favor of them—but I have quite a lot against shelling nuts. As a matter of personal prejudice and sloth, this seems to me to be tedious, mostly indoor work, as aggravating as fixing wobbly chairs and no more interesting than painting window sashes. Fortunately it is possible around here to have your nuts without personally cracking them. There are lots of productive trees in the mountains, and harvesting them was once a semicommercial proposition. There used to be a tradition that if you had enough kids you could pay the fall taxes by setting them to work shelling nuts to sell in local markets. Taxes and kids both having become more formidable, this is now a memory instead of an economic fact. Assiduous social foraging, however, will turn up a few clans with whom it is still possible to trade for jars of butternut, hickory nut and walnut meats.

Besides the obvious advantages, not doing nut work keeps a forager fresh for pawpaw season, which comes at about the same time as nut season. The pawpaw is a wild exotic, a relative of, and almost as good as, the tropical custard apple. Here we are almost at the northern limits of the pawpaw's range, and it is found chiefly in patches on bottomlands. Once—and perhaps still—along the St. Joseph it was called the Michigan banana. How and why it got this far is a mystery, but a blessed one.

The approved way of finding pawpaws is to organize your life so you have an absolutely free day in the best part of September. Take to the Potomac, Monocacy or Susquehanna River in a canoe. Put in a congenial friend or two and several bottles of dry white wine. Paddles may be carried but should not be overused, the proper method of progress being a slow, idle float. In God's good time you may come upon some clumps of small, slender trees with big, jungly-looking leaves. As you near them you will get a whiff of a sweet jungly smell. This will be the pawpaw patch. Shake the trees *very gently*. Some elongated, potato-shaped fruits will fall to the ground. Pick up the ones that give a little splat as they hit. Eat as many as you can and drink the wine.

Green pawpaws are rock hard and bitter. Overripe ones disintegrate into a kind of sticky compost. Pawpaws are exactly right—and great for a few days—when they resemble old, bad bananas in color and consistency. In this condition pawpaws are difficult to transport and ugly to look at, which is why they are seldom items of commerce. But every rightminded possum, raccoon, fox and squirrel in the neighborhood, as well as a great array of birds and insects, as well as we foragers, knows when pawpaws are ripe and admires them. Thus there is considerable urgency in getting to the trees before the beasts have stripped them. Immediate sensual rewards aside, making the right pawpaw connection gives a satisfying, attuned-with-the-world feeling of being in exactly the right place at precisely the right time.

Like winning football quarterbacks, successful foragers tend to take what is given them—that is, they are opportunists rather than grand strategists. But toward the end of that remarkable year, I decided to force matters, to search intentionally for something I had not had in a long time but remembered from the St. Joe River days as being worthy—persimmons.

The persimmon tree, like the pawpaw, is essentially a tropical species (related to the Asiatic ebonies) and probably never was common in the central Pennsylvania mountains. It is increasingly rare now because many of the larger trees have been cut, the fine-grained wood being useful for certain kinds of cabinetmaking. Naturally, those who know the whereabouts of surviving trees are chary about passing the information on to those who do not. But late last fall, after some traipsing and trespassing, we came upon a grove of four substantial and bountiful trees. They were on land owned by a congenial octogenarian who said he had been happily feeding in this grove for 70 years and at this time of life did not mind sharing, moderately, with serious-minded colleagues.

As with pawpaws, getting persimmon fruit at the right time is critical. A green persimmon (actually yellow) is astringent beyond belief, puckering the mouth like no lemon or pickle ever invented. They are dead ripe when they are about the color of badly frozen apples and the consistency (as well as shape and size) of stewed prunes. Persimmons being very slow ripeners, this seldom occurs until Thanksgiving or later. They will continue hanging on bare, leafless limbs until they are knocked off by winter winds and blizzards. Persimmons, therefore, are among the very last of the season's forage.

After one has found properly ripened persimmons, one first eats a few handfuls. Persimmons are sometimes called sugarplum trees, which may give you an idea of their taste. The second thing to do is pick up or pry out of the frozen ground all the windfall and get them back to the kitchen immediately. Mash them and strain out the seeds and skins. Get at least two cups of the pulp. Mix up ¾ of a pound of butter with two cups of honey and three jiggers of maple syrup. Someplace else mix four cups of flour, four eggs and a tablespoon of baking soda. Then put those two batches together and add the two cups of persimmon pulp and a cup or so of wild walnut or hickory nut meats. Stir some more. Put the batter in greased cake pans and bake for an hour, or until ready, at about 325°. When it goes into the oven the batter is orange, taking its color from the persimmon pulp. While baking, it turns a deep, rich chocolate color. It also turns magnificent, the finished loaves being a bit stiffer than pudding, much juicier than any cake and, I think any reasonable glutton would agree, better than both. In a year of exceptionally keen competition, what with sulderade and all those oyster mushrooms, Persimmon Delight was the best of all. The loaves can be frozen and the Delight judiciously and therapeutically spread out through the dark winter.

There is something else that can and will be said about foraging. I will approach it by analogy. The recognized function of a photo album is to serve as an aid to nostalgia. A loaf of Persimmon Delight can perform much the same function, although it works on the mind through the taste buds rather than by means of the optic nerves. In it is the taste of December persimmons, October walnuts, August honey and March maple. It does not taste like these months or seasons any more than a flat photographic print of, say, a bride really looks like a bride. However, the tastes are indelibly associated with phenomena that cannot be reproduced, inasmuch as they occurred in the past and will never reoccur in the same way—certain peculiar conditions of sun and snow, sensations of fire and beesting, how daughters and paramours once sounded and looked in summer fields. We seldom think of them as having that function, but our gustative and tactile faculties stimulate recall in very subtle and moving ways.

Also, of course, something like Persimmon Delight permits you to have your memories and eat them, too.

Conceits and Feats

When it comes to things like acid rain, vine-ripened tomatoes and shoplifting, we appear to be backsliding rather than getting ahead. On the other hand it remains a matter of general faith and some comfort that athletically at least we seem to be on the steady improve. In track and field, for example, performances (or at least record ones) in all events are about 20 percent better than those of a century ago and the same progress is evident in sports that are similarly measurable. However, the statistical improvement of records has given rise to an odd, common assumption—that athletically we are innately superior to our ancestors and can do sporty things that would have been physically impossible for them to do. This notion cannot be proven; to the contrary, both logic and available historical evidence indicate that the athletic interests and activities of, say, nineteenth- and twentieth-century America are so different that they are as incomparable as apples and oranges. Furthermore, where some rough comparisons can be made there is little to suggest that things have changed greatly so far as natural ability is concerned.

One obvious and influential difference is that we take athletics much more seriously than our forebears did and in consequence have devoted a lot more effort than they did to devising equipment, facilities, training, and competitive aids that improve athletic efficiency. To a considerable, if immeasurable degree, improvement in record performances must reflect technological progress rather than an increase in physical prowess.

By way of speculative example. Foot racing, especially sprinting, is one of the few sports whose popularity (and basic rules) have remained fairly constant in this country for the past two hundred years. Throughout the nineteenth century, the sporting *pièce de résistance* of many civic celebrations and holidays was a hundred (or approximately so) yard dash. Long before anybody had thought of or wanted to bother with spikes, starting blocks, plastic or even cinder tracks, it was the custom to mark out a course in the main street, assemble a sizable field behind a line drawn in the dirt, and send it off with a volley of rifle fire. There were thousands of such contests and they may have included some extraordinary performances but it is impossible to be certain since these were invariably match races pitting local champions. The main interest was in who won, not how fast they ran. (Without question, we are now more avid athletic clerks and archivists and this too has a bearing on the quality and certainly quantity of our records.)

By sifting through contemporary reports, it seems likely that a man who could run a hundred in less than 11 seconds was considered a real burner during most of the nineteenth century. (In 1876, the first year in which a national championship was staged, the dash was won by F. C. Sapporta in :10.5.) Such times would embarrass a high school sprinter these days. On the other hand, if they or even a Carl Lewis competed under the conditions which generally prevailed a century and more ago—that is, came to the race directly from work, took a few belts of raw whiskey as a training aid, ran in clodhoppers or barefooted down a track rutted by wagons and pitted by stock—the chances are their times would be much closer to those of the earlier flyers than to our current records.

In the nineteenth century, society was much more rural, labor-intensive, and physical than it is now. Contemporary work generally being more sedentary and less exhausting, many of us are able, eager, and perhaps need to engage in hard sport during our free time. In the past when there was proportionately a lot more muscle work, the general and reasonable opinion was that a good thing to do with leisure was sit down in a cool, soft place, eat and drink heartily. (This still tends to be the case, joggers, racquetball clubs, and health spas being fairly rare in our remaining rural areas and blue-collar communities.)

A nineteenth-century farmer or cowboy, lumberjack or freight hauler might enjoy coming into town for a Fourth of July footrace but would not have had much inclination or leftover juice for spending a lot of time training for his event. Even if he had, he probably could not have afforded to do so. Professional and even serious amateur athletes were almost nonexistent in America before the Civil War and rare, based mostly on the East Coast, during the rest of the century.

By all accounts, when earlier Americans did have the time and were in the mood for games they became as competitive, emotional, and pugnacious about them as we do. However, they did not see much sense in devoting a lot of effort and money to getting ready for them, either athletically or organizationally. Popular sporting events were apt to be impromptu and much more literally pastimes than is now the case. It is instructive to remember that a great many social gatherings were once commonly called frolics—weddings, political rallies, corn huskings, goose pullings (a live goose with a greased neck was hung upside down from a pole and men on horseback competed to see who could first wrench off the bird's head). Recreational eye gouging, ear biting, stomach stomping, and butcher knife duels were also popular in frontier America. Rather than being occasions for calling out the police or military, they were commonly regarded as good if rugged sports and fine spectator events. For better or worse many of these pastimes have disappeared and there is absolutely no way of matching performances in them against those in the events that are now contested.

What might be called work games, competitions involving on-the-job skills and tools, were probably taken more seriously than the frolic sports. From the Atlantic coast to the Great Plains, rail splitting, for example, was a universal labor. Depending on the type and condition of the timber, making 100 or 150 ten-foot rails during a day was considered an all-star performance. Inevitably competitions occurred and reputations were made in them. The renown that one lanky young Illinois frontiersman achieved because of his good moves with a maul and wedges started him in a career that eventually led to the White House.

Breaking up rocks to surface roads occupied thousands of men during the first half of the nineteenth century. The largest and most continuing stone-crushing effort involved the construction of the National Road—the first federal highway, which, after forty years of work beginning in 1806, was extended from Baltimore to Vandalia, Illinois. Early in the 1820s when the Road was being cut through the Allegheny front in southwestern Pennsylvania, there occurred a competition that was remembered for years afterward. Robert McDaniel of Fayette County, Pennsylvania, was recognized as the champion rock breaker on the gangs working in that section. One day a challenger, Elias Gilmore, identified in surviving reports only as "one of the most vigorous men on The Road," showed up to test McDaniel. Crews knocked off work and gathered around to watch as the two men started at dawn making little ones out of big ones. (To pass federal inspection rocks had to be broken in small enough pieces to pass through an iron ring seven inches in diameter. Making enough of them to cover eighty feet of roadbed was considered to be a good, noncompetitive day's work.) Stripped to the waist, without any sort of break the two men hammered away through the lunch hour. By midafternoon McDaniel was exhausted and, as the report went, "threw down his hammer and granted the palm." Gilmore continued swinging until dark and crushed enough stone to cover a 150-by-50-foot piece of roadbed.

Though they may not have been of the class of Abraham Lincoln, Gilmore, and McDaniel, it seems likely that in their day there were hundreds of men, work-athletes, who could break rock and split rails and do many other strenuous jobs better and faster than anyone can today. Their accomplishments do not prove they were better men than we are, just ones with much different interests and acquired skills. Likewise they cannot be regarded as our inferiors because there was not a jack among them who ever executed an under-the-armpit, behind-the-head, in-your-face slam dunk.

Here and there are surviving stories—if not records of the contemporary sort—which are thought-provoking to say the least, since they involved activities very similar to those which we are still sportily pursuing. There was, for example, a Kentucky frontiersman and Indian fighter by the name of Robert McClellan. One day, probably in about 1815, he was frolicking with some spirited friends in Louisville and challenged to jump over a double team of oxen. Accepting, McClellan is reported to have said, "I feel like a colt. Now boys look at me." Whereupon he sailed over the team, making, it would seem, a long jump of at least twenty-five feet and one which could have been very close to Bob Beamon territory. There is a cryptic footnote in the account of this happening. "Concerning his [McClellan's] marvelous jumping abilities it is said that he was excelled only by one William Kennan, a noted scout of the border." No further mention is made of the super-marvelous Kennan.

A U.S. Cavalry captain, Sylvanus Cushing, was regarded as a "giant in strength," being over six feet tall and weighing 220 pounds. One day Cushing was out hunting with a party that included Buffalo Bill Cody. Cody's horse fell so that it could not move and Cody was spreadeagled underneath, his legs extending on each side of the horse's belly. Cushing dismounted, stooped over, "took the forelegs in his hands and bracing himself against the withers, gave a quick heave and rolled the animal clear over the saddle and off Bill." All of which would seem to have approximated a clean jerk of about five hundred pounds.

John Colter was one of the first of the American fur trappers and mountain men in the Rockies. In 1809 he was captured by the Blackfeet and was given a choice of either being tortured at the campfire or running—after being stripped naked—for his life. He took the latter option and there followed a lethal super-marathon. Exceedingly well motivated, Colter burned the first five miles, outdistancing all but one of the five hundred pursuing warriors. With a quick martial-arts move he tumbled this man, grabbed his spear, and killed him with it. Then Colter dived into the Jefferson River, swam and floundered some thirty miles down it, hiding under driftwood rafts to escape the vengeful Blackfeet who were ranging the banks. That night Colter got out of the water and during the next six days, without shoes or food except for a few roots and berries, ran three hundred miles across some very rough, thorny sections of present Montana and Wyoming. He eventually reached the nearest fur trading post safely. Colter was the only witness to all of this but circumstantial evidence convinced contemporaries that his story was, in the main, true.

It seems very likely—because they were so often cited—that men like McClellan, Cushing, and Colter turned in extraordinary performances, but the details of what they did are too sketchy to be translated into hard statistics. However, there was at least one astonishing nineteenth-century feat performed purely for sport and precisely recorded by reputable men. It took place in the spring of 1877 in western Nebraska but has since been almost forgotten, certainly by modern record keepers. Yet if what is said to have occurred did occur, the happening requires some drastic rethinking about the particular event, the mile run, and the history of athletic progress in general. The circumstances—of some sporting interest in themselves—leading up to this phenomenon are as follows:

In the mid 1850s, as teenagers, Frank and Luther North emigrated from Ohio with their family to the Platte River valley. Immediately the brothers struck out on their own, turning their hands to a number of jobs in the upper Great Plains. Variously they were cowboys, farmhands, freight haulers to the Colorado gold fields, Indian traders, professional wolf and buffalo hunters. Eventually they became ranchers and prominent men in Nebraska with the means and leisure to be, so far as the frontier environment permitted, bon vivants. They were acquainted in the saloons and gambling halls of the region but also were accepted by and had a taste for the most respectable society. They shot billiards, tossed "Quotes" and rolled tenpins with bankers, judges, and military officers; played genteel games of croquet and lawn bowls with their daughters and wives. Not infrequently they dressed themselves up to the nines and took the steam cars to Chicago, where they put up in good hotels, dined and wined in fine restaurants, attended the theater and even opera.

These sophisticated interests aside, the Norths were renowned as exceptionally tough and daring frontiersmen. Among other talents, the brothers were fine marksmen and Frank was reputed to be "the best revolver shot in the west." According to Luther, his brother, a man named John Talbot, and Wild Bill Hickok, the gambler and sometimes peace officer, would often meet at Talbot's roadhouse near Cheyenne, Wyoming, for competitive shootouts. "Frank would nearly always win," recalled Luther, "with Talbot second and Wild Bill third. I never saw Wild Bill shoot with his left hand either, although he was always called a two-gun man, and as to shooting from the hip, I never did see a man shoot from the hip, though I have seen such gunmen as Wild Bill, Jack Hays, Doc Middleton, Joe Hall and others."

The reminiscences of the Norths give the impression that they regarded Hickok as a fairly peculiar fellow whose abilities and, particularly, courage were vastly overpublicized—an opinion that has now become general among western historians. They thought much better of another sometime crony (and partner in the cattle business), Buffalo Bill Cody. Though Cody arrived too late to have much genuine frontier experience and was essentially a trick, exhibition shootist, the North brothers both thought he was the best marksman they knew when it came to firing a rifle from the saddle of a running horse.

During their knockabout youth, the Norths had become well acquainted with a number of Pawnee Indians, learned their language, and developed a close rapport with the tribe. The Pawnees were a small but pugnacious nation whose historical misfortune it had been to have their homelands located along the Platte, directly in the main route that whites chose to use during their emigrations to Oregon and California. During the 1840s the covered wagon pioneers regarded them as the worst scourges on the plains, but by the time the Norths settled in Nebraska the tribe had for the most part been confined to a reservation on the Loup River near the North's homeplace. Their relations with the settlers were fairly amicable—because they no longer had the resources to harass the whites and because they were willing to cooperate with whites as a means of defending themselves against and continuing to raid the Sioux and Cheyennes, larger plains tribes who had been bitter enemies of the Pawnees for generations.

In 1864 the Brule and Oglala Sioux went to war, striking suddenly at a number of isolated settlements and military posts in the northern plains. In the course of things they made a passing swipe at their traditional foes, the Pawnees, killing eight members of the tribe whom they caught near the Platte. So far as whites were concerned, defense and retaliation were complicated by the fact that the federal military was then largely preoccupied with events occurring in northern Virginia. In consequence, a company of one hundred Pawnee braves were recruited to serve as scouts under the command of their old friends, the Norths.

The Pawnee light cavalry did good work, particularly in August 1867 at the battle of Plum Creek, where in a gaudy twenty-mile running battle they defeated a sizable Cheyenne war party. Through the remainder of the 1860s they were kept in service, principally to protect the transcontinental railroad labor battalions. Frank and Luther North (respectively a major and captain), the gentlemen adventurers who led them, became considerable national celebrities, at least as well known as Hickok and Cody and in the main regarded as more substantial frontiersmen.

By the early 1870s floods of new, would-be settlers were arriving in Nebraska and to satisfy their demand for land, the Pawnee tribe was unceremoniously removed from its Loup River reservation and packed off to a new, very poor and barren one in the Indian Territory (present Oklahoma).

In 1876, feeling themselves generally abused and particularly wrathy about an influx of gold prospectors and miners into their traditional Black Hills hunting grounds, the Sioux and Cheyennes rose once again. By the Fourth of July they had rubbed out the flamboyant George Armstrong Custer and all of his troopers.

Immediately upon receiving this news, General Philip Sheridan, the doughty hero of the Valley of Virginia, who was then commanding military operations in the trans-Mississippi West, set about organizing a campaign to avenge Custer and settle the hash of the plains hostiles once and for all. Late in the summer, Sheridan ordered Frank North, technically a civilian, to proceed at once to Indian Territory with his brother Luther and there find and recruit one hundred Pawnee Scouts. They then were to ride to Fort Sidney, near the Wyoming-Nebraska line, where they would support General George Crook in a winter campaign against Dull Knife, the most important Cheyenne war chief still on the loose.

In September the Norths arrived in Indian Territory, at Fort Sill, which was a cavalry post and the headquarters for the Wichita Indian reservation but lay more than one hundred miles from the unpromising compound into which the Pawnees had recently been herded. However, rumors had already reached the latter tribe about their possible assignment and they were delighted by the prospect. Not only were they eager to get into the field against the Sioux and Cheyennes, but in Indian Territory they had fallen on very hard times. They were homesick for Nebraska, impoverished and starving. Any enterprise that would take at least some of them north and give them regular rations was a welcome one.

To find out what might be expected, the Pawnees sent Koot-tahwe-coots-oo lel-le-hoo La Shar (called Big Hawk Chief by the whites) to Fort Sill to make medicine with Frank and Luther North. Big Hawk Chief was then twenty-three years old and was renowned as the strongest and swiftest runner among all of the Pawnees, a tribe that had a great reputation in the plains for producing speedy warriors. The lone messenger covered the 120 miles between the Pawnee encampment and Fort Sill in twenty-four hours. His run so impressed the chief of the Wichitas that he asked if Big Hawk Chief thought he could ever duplicate it. The Pawnee said he could and would two days later, when he expected to return to his people with messages from the North brothers. The day after, accompanied by the doubting Wichita chief, who was mounted on a horse, Big Hawk Chief started running. At the 60-mile mark the horse went lame and before a remount could be found, the runner was long gone. The Wichita was finally able to continue and arrived long after dark to find Big Hawk Chief lounging in his lodge. On the return trip he had improved his time to twenty hours. If the distances and times were anywhere near those estimated, the Pawnee had run the equivalent of nine back-to-back marathons during a three-day period.

The Norths finally rounded up their Scouts, found horses for them, and in midfall started toward Wyoming. On Thanksgiving Day of 1876, Crook's troopers, the Norths, and the Scouts cornered Dull Knife (with four hundred of his warriors and about one thousand women and children) encamped on the Red Fork of the Powder River at a place where the stream cuts a passage through the Big Horn Mountains. It was a bitter, bloody day. Gritty, stinging snow squalls filled the air and at dawn when the fighting began the thermometer stood, as Luther North remembered it, at 25° below zero. "We fought them on various parts of that field," Luther North later wrote, "and I have always thought they put up a wonderful fight. The Cheyenne ran up among the rocks north and west of the village and we never drove them a foot further."

Luther North and a single Pawnee sergeant, Pwiakke La Shar, under a hail of covering fire, worked their way into the hostiles' camp and stampeded their horses. Without reserve mounts, having very little food or ammunition left, the Cheyennes were in a hopeless position. That night, during which a foot of snow fell, they slipped out of the rocks from where they had fought so valiantly and limped westward into the Big Horns, the surviving warriors covering the retreat of the women and children. Luther North and four Scouts, including Big Hawk Chief, were detached to follow them, but the Cheyennes had neither the strength nor the means to continue fighting and surrendered shortly thereafter.

The Norths and the Pawnees remained in the field until midwinter of 1877 and then returned to Fort Sidney. They stayed there for the next three months, until the Scouts were mustered out for the last time. (They were sent back to Indian Territory with their army pay and, as a bonus, two hundred head of good horses captured from the Cheyennes. Frank North rode with them to keep them from being despoiled by whites.)

Sidney, both the fort and the civilian community that had grown up around it, was perhaps the wildest western town then booming. It was full of troopers celebrating the end of Indian wars and was also a railroad station–supply depot for the gold fields of the Black Hills. Luther North vividly recalled the style of the place. "The town was full of saloons and dance halls, and gamblers and gunmen were very plentiful. . . . I was in a saloon one night when a gambler and a teamster quarreled. They pulled their guns and commenced shooting at each other. They were at opposite ends of a billiard table, and one of them fired six and the other five shots. The gambler was hit once, losing the lower tip of his ear, and the other man wasn't touched. . . . The two men were considered fairly good shots . . . ," North concluded with some irony, "but it shows that all gunmen were not so deadly as some writers would have us believe."

The Norths wisely kept their Pawnees under tight discipline and out of the worst Sidney hellholes. By way of diversion and to entertain some visiting eastern businessmen and politicians, a sham cavalry battle was arranged between the Scouts and H Troop of the Third Cavalry. It was judged to be a draw but a fine attraction with a lot of whooping, daring horsemanship, and harmless gunplay. Later a baseball game was organized in which the cavalry did defeat the Indians.

One day in April, Luther North began telling some stories about the footspeed of his man, Big Hawk Chief, who had previously been examined by the Fort Sidney surgeon, who "stripped [him] and went carefully over him, stating afterward that Big Hawk Chief was the most perfect specimen of man he had ever seen."

An acquaintance who was never after named but was probably an army officer apparently said something to the effect that seeing was believing. Thereupon North called up Big Hawk Chief and with his friend went off to watch the Indian run a mile on the horse-racing track that engineers had laid off on the prairie outside the fort. It was a half-mile course, presumedly a soft dirt one, chewed up by the ponies for whom it had been built. Luther North brought along a watch that at least had a second hand and may have been one of the one-fifth-second stopwatches then being used to time horses and dogs. (There was also a pack of thirteen racing greyhounds at the fort.) In the critical matter of accuracy nothing is now known about this timepiece.

At the track Big Hawk Chief stripped down to his breechclout and started running at the crack of a revolver shot. He cruised the first half mile, according to North's watch, in 2:00, then dug in and came back with a 1:58 for his second half split. North and his friend were so amazed by the clocking that they immediately sent for a steel tape and remeasured the track. They found it to be exactly a half mile and that Big Hawk Chief had therefore run a full mile in 3:58.

The proud Luther North later said of his Pawnee flyer, "He had no equal. I would like to have seen him matched against the best runners in this or any other country." (Any such contest might have been fairly one-sided. A few months later in New York, at the U.S. championships of 1877, the mile was won by R. Morgan in 4:49¾. Nearly twenty years later the first Olympic winner in the fifteen hundred meters was Edwin Flack of Great Britain, whose time was 4:33.)

The first, almost reflexive reaction is that Big Hawk Chief's time was impossible and his run unbelievable. We *know* that the "magic" four-minute-mile barrier was not broken until 1954 when Roger Bannister ran a 3:59.4 in what was then and still is considered one of the greatest record-breaking performances of the twentieth or any century. The truth is that, of course, we do not know this, only believe it.

Big Hawk Chief's time was not and cannot be certified as records now are. All sorts of technical questions can be raised about the distance, start, and the watch. Also, Luther North may have been a flat-out liar, though this was not his reputation and there were few men in his day who would have cared to so name him. On the other hand, we do not *know* that Big Hawk Chief did *not* run 3:58 on that dirt racetrack, just as Luther North observed and solemnly reported that he had. For that matter, we do not know that this was the Pawnee's personal best effort, nor even that some or many other extraordinary men before had not run faster than he did.

The moral is, perhaps—Beware of chronological chauvinism.

The Great American Creeple

The classical monsters—harpies, hydras, scaly giants and werewolves—who tore down mead halls, carried off virgins and in general made a plague of themselves, apparently suffered from seasickness. At least they never crossed the Atlantic. Americans have had to deal only with an occasional Jersey Devil or Sasquatch, and these local abominations have been pretty much the bumpkin type and have kept to the outer thickets.

Still, a society benefits from at least one formidable fright. What would Old England have been like without dragons, or Rumania without vampires? Lacking exotic imports, Americans have created their own monster. The Old Country technique was to begin with the tales of minstrels, poets and magicians, embroidering on these until something suitably frightening to children was created. Being down-to-earth people, Americans started with a real beast, and then talked it up to a level where it could hold its own against any demon St. George ever knew or slew.

Our monster is the rattlesnake, which European-Americans first encountered four centuries ago. Since then we have been so obsessed with the reptile, told so many tales about it that few of us can distinguish the real beast from the mythic monster.

The first news of the rattlesnake may have been taken back to the Old World by a Spaniard, Pedro Cieza de León, who in 1554 published an account of his travels in Peru. Considering the tales that would follow shortly, his report was matter-of-fact: "There are other snakes which make a noise when they walk like the sound of bells. If these snakes bite a man they kill him."

In 1630 a New England divine, the Rev. Francis Higgeson, wrote, "Yea, there are some serpents called Rattle Snakes that have Rattles in their Tayles, that will not flye from a man as others will, but will flye upon him and sting him so mortally that hee will dye within a quarter of an houre, except the partie stinged have about him some of the root of an Herbe called Snakeweed to bite on, and then he shall receive no harme."

Thomas Morton, writing home to England in 1637, commented, "There is a longe creeple that hath a rattle in his tayle, that does discover his age; for so many years as hee hath lived, so many joynts are in that rattle, which soundeth like pease in a bladder, and this beast is called a rattlesnake."

It would be unseemly to continue further without noting that the brief quotations from Cieza de León, Higgeson and Morton, as well as some of the intriguing information that follows, appear in a marvelous work of science and art, *Rattlesnakes: Their Habits, Life Histories, and Influence on Mankind*, written by the late Laurence M. Klauber. Klauber was an engineer-turned-herpetologist who was the consulting curator of reptiles at the San Diego Zoo for more than 40 years. His *Rattlesnakes* touches on everything from the morphology to the mythology of the reptiles, and is without rival as the standard reference work on the subject. So complete and exhaustive is it that citing Klauber whenever material from his study is used would be tedious, but it would also be misleading and ungrateful to minimize the work as a source. The views of other authorities and numerous personal observations of rattlesnakes follow, but almost anything anybody wants to know about rattlesnakes appears in Klauber's two-volume, 1,500-page masterwork (University of California Press).

After the first more or less straightforward accounts of the American creeple, stories about rattlesnakes began to blossom in luxuriant fashion. In 1642 Thomas Lechford supplied the comforting information that, if bitten, a man would turn the same color as the offending snake, i.e., "blew, white, and greene spotted." Since this color combination does not correspond to any known species of rattler—or man, either—Lechford was perhaps the first in a long line of reporters contributing to rattlesnake lore without ever having seen one of the reptiles. In any event, after Lechford the mythologizing of the creeple began with a vengeance.

A cardinal rule of monster-making is to think big. Some of the bigger and better rattlers reported since (and catalogued by state) include:

Arizona. Early in this century a 14-footer lived in the Huachuca Mountains along the Mexican border. It had the unpleasant habit of slithering after prospectors, running them into their cabins and laying siege to them. The size and behavior of this snake suggest it may have been a descendant of the Apache snakes that once infested this area. Apache shamans would talk giant snakes into ambushing whites.

Arkansas. A rattler of undetermined length but with a head the size of a water bucket and rattles as large as a coffee cup.

California. In 1881 a rattlesnake attacked a horse pulling a rancher's wagon. The rattler dragged the horse and wagon to the edge of a gulch and there tethered the horse by hitching itself around the animal's leg and a tree. For extra purchase the snake drove its fangs into the tree trunk. The astonished rancher finally gained the upper hand by shooting the reptile; he said that it measured 12 feet.

Florida. Eighteen feet.

Ohio. In 1808 a 12-foot rattler measured 15 inches "around the shoulders."

Oklahoma. In the 1950s a 21-foot rattler was killed near Poteau after it had bitten a lady in the foot.

Perhaps the largest rattler of all was seen in Baja California. Klauber received a letter from a rancher's wife describing an encounter between a boy and this snake, which was between 25 and 30 feet long and "hissed with such force it sounded like a bull." A rattler of that length would have weighed close to 400 pounds.

According to verified records, the eastern diamondback rattler (*Crotalus adamanteus*) is the largest of the species. The longest Klauber personally measured was a bit over six feet, but there are dependable records of seven-footers. The longest diamondback for which Klauber believed there was indisputable evidence was an 8'1½" specimen.

Ross Allen, the famous snake man of Silver Springs, Fla., put monster snake stories into perspective when he wrote in 1953: "I have been in business for nearly 28 years, during which time I have received from 1,000 to 5,000 *Crotalus adamanteus* annually, a total of about 50,000 altogether. The largest specimen I personally measured was 7'3" in total length, exclusive of rattle, and weighed 15 pounds. For years I offered a reward of $100 to anyone who brought in an eight-foot Florida diamondback, dead or alive. In recent years I have offered $200, without results."

Spoilsports such as Klauber, Allen and their colleagues have tended to exert a depressing effect on tellers of long-rattlesnake stories. However, when it comes to behavior, folk reporters have not been hampered by tape measures. Rattlesnakes have been said to possess the following characteristics and capabilities:

– They bite with their fangs but inject venom through forked tongues.

– The breath of rattlesnakes is poisonous, and there is a dust in their rattles that is lethal if inhaled.

– By staring at their intended prey, including men, rattlesnakes exert a hypnotic influence, immobilizing victims so that they can be killed; some are killed by the evil eye. Many people have reported seeing small animals, mostly birds and rabbits, so entranced that they walked into the jaws of a snake. There is a theory that a rattler must kill by hypnosis or something similar because if it killed by injecting venom, then the rattler itself would die eating the poisoned prey.

– Rattlers will strike and eject poison into a spring, then hang about waiting for some unwary creature to drink the water, after which it dies and becomes snake food. In the same fashion, rattlers will strike and contaminate holly trees, then wait for dead birds that have eaten the berries to fall down around them.

– Rattlesnakes will swallow their young, giving the brood temporary refuge from danger.

– Rattlesnakes milk cows. (Among others, Sherlock Holmes and Mowgli reported the fondness of snakes for warm milk.)

– Rattlesnakes mate for life and are inseparable. If one is molested, the other's anger and desire for vengeance is intense. Thus, after a member of a covered-wagon train killed a female rattler on the Kansas prairie one morning in 1853, the male trailed the wagons all that day and night, finally killing one of the emigrants.

– If a snake bites a man and the snake dies, the man will live. If the

man dies, the snake will live. (A variant has it that if the snake is killed before the fate of the man is determined, the toxicity of the injected venom will be increased.)
– If a snake bites a nursing mother, her infant will be poisoned. However, in the South, the last bastion of gallantry, rattlesnakes will not bite women. In other conservative areas a snake will bite some women but never a good (in the severely technical sense) woman.
– Rattlesnakes will not bite a person wearing an anklet of snake bones. White ash, onions, burning shoes, king snake oil, tobacco, urine and hair balls of a cat will keep away rattlesnakes, depending on what part of the country you are in.
– A motorist traveling in the Salton Sea area of southeastern California suffered a puncture. He lacked a spare tire but understood the ways of the desert. He backed over a nearby rattler, which struck the tire. The rubber immediately swelled from the effect of the venom and the puncture was sealed.
– If a rattler strikes an ax head, the metal will become discolored and the infected piece will break. On the other hand, if a rattler strikes an ax handle, it will swell until the head is popped off the shaft. If a rattler bites a piece of meat, the meat will turn green.
– If a bee stings a rattler, the bee will soak up venom that will not hurt it but will be fatal to anything the bee stings.
– On the useful side, puree of rattlesnake, rattles and venom is good for toothaches, tuberculosis, chills, fevers, rheumatism, ringworm, deafness, bad complexion and acid indigestion.

These claims made against and for rattlesnakes are selected from among hundreds available. They all have been proven false, but during the course of 35 years of talking about rattlers with a variety of people, I have found somebody who believes each one. In some cases *suspects* is perhaps a more accurate word than *believes*. A cowboy will say, for example, that it seems improbable that a rattler will poison water but he has heard good secondhand reports of this happening and that until he has better information he is not about to drink out of Yaqui Springs, which is a notorious rattler hangout.

By no means are all of the believers ignorant backwoodsmen. In fact, the farther removed people are from the company of rattlers the more inclined they are to be gullible about them. I know of one corporate lawyer who believes (because he was told it was so as a boy by the family gardener) that the fore rattle of a snake contains dust that will cause blindness. On the other hand, there is a Tennessee moonshiner, a qualified backwoodsman, who has heard this evil dust story, knows that it is an old wives' tale and will prove it by grinding up a rattle and sprinkling the powder on his food. However, this man also pulverizes rattles, mixes the powder with a quart of raw moonshine and spreads the mixture around his still when he is working there. He does so because his granddaddy told him this was a good way to keep down rattlers.

Beyond what might be called rattlesnake unnatural history—that is, information that somebody once or still believes to be true—there is an abundance of rattlesnake whoppers. These are tall tales that not many believe but that provide good yarns. One of these is sufficient to suggest the flavor of many. The story is told in the Seven Mountains of Pennsylvania that a man set out to build a cabin. In dressing the first log he disturbed a rattler, which sank its fangs into the timber. The log began to swell from the venom and continued to swell until it provided enough lumber for a 12-room house. When the building was finished, the man painted it, forgetting the well-known properties of turpentine for neutralizing rattler poison. After the first coat of paint the house began to shrink and eventually was reduced to the original log.

Nothing could be so clever, diabolical and stout as the mythical rattlesnake. Nevertheless, real, flesh-and-blood rattlers are remarkably interesting and indeed formidable. They belong to a family known as the pit vipers, which includes two other venomous American snakes, the copperhead and cottonmouth. Pit vipers are thought to have originated in the Mexican highlands and to have moved north and south.

There are 15 known species of native rattlesnakes. (A rattlesnake is any pit viper with a rattle.) Though these seem to do best in dry warm country, they are found all the way to the Canadian border (and even beyond) in a variety of habitats: woodland, swamp, coastal, prairie, rocks and sand. The optimum temperature range for rattlers is 80° to 90°. Being cold-blooded, they cannot survive ground temperatures above 105° to 110°. Therefore in very hot weather they usually go underground during midday and hunt by night. Below 55°, rattlers become inactive and they may die in freezing weather so they seek hibernation dens in the fall and remain there until spring.

Given their catholic taste in habitat, rattlesnakes have existed at one time or another almost everywhere in the U.S. There are now a few localities where rattlers have been exterminated (Delaware and Maine, for instance).

Females mate when they are two or three years old and bear their young in broods of a dozen or so; thereafter they show little maternal concern. In myth, great age is often attributed to rattlers, but in fact a 10-year-old has led a full life. There are reports of individuals living 20 years, but this is exceedingly rare. The notion that the number of rattles on the tail indicates the number of years the beast has lived is false. A rattle is added each time the snake sheds its skin, which in young animals may occur three or four times a year. Also, longer strings of rattles usually break off because they are brittle. Reports of huge rattle strings, containing 40 to 80 segments, are often heard. However, it is easy to fabricate long chains by taking the rattles of many snakes and fastening them together.

Rattlesnakes are predators, never taking vegetable food except by accident. They prey on small mammals up to the size of rabbits, on birds, lizards, amphibians, occasionally on fish and other snakes. Rattlers are heavy-bodied and not especially agile. At top speed, which they can maintain only briefly, they crawl at a rate of about three miles per hour, walking pace for a man. They have fairly good vision but are shortsighted, not able to perceive movement much more than 15 feet away. There is some dispute on the matter but it is generally thought that rattlers, which are earless, do not hear in the conventional sense. However, they are sensitive to vibrations of the earth. Rattlers are good swimmers and are not timid about entering the water.

A variety of mammals will prey on rattlers, as will predatory birds, most especially the red-tailed hawk. Wild turkeys, domestic turkeys and chickens will sometimes kill rattlers. The king snake has a great and deserved reputation as a rattler killer, but this is not an obsession or a matter of natural law and order. If a king snake comes upon a smaller rattler it may kill and eat it. Cannibalism among rattlers is rare but not unknown.

In evolutionary terms, the rattlers are a comparatively modern group of snakes, possessing a variety of sophisticated and newfangled adaptations that their relatives lack. For example, there are the pits that give this branch of the vipers its family name. These appear on each side of the head between eye and nostril. They are heat sensors and enable the snake, at distances up to a foot, to detect a temperature difference of as little as one degree centigrade higher or lower than that of the background. These sensors are particularly useful to the snake as it hunts in dark tunnels.

The rattle is an interlocking series of horny, multilobed segments attached to the tip of the tail. When the segments are "rattled" against one another as a result of an intentional muscular movement by the snake, they produce a distinctive buzzing sound. The rattling apparatus is unique, and one wonders why it evolved and what its function is. It has been suggested that rattlers rattle as a kind of sporting gesture to give other creatures a head start before the snake comes after them with the intent to poison or mesmerize. It has also been claimed that the rattle is used in courtship, as a device for communication, or for attracting curious and potential prey. These theories have now been discarded, studies seeming to show that a snake rattles to bluff away creatures that might do it harm. "Don't tread on me," is apparently the message and function. The belief that rattlers will always rattle before striking is false. Whether or not a rattler makes a noise depends upon the circumstances and, very likely, the mood of the snake.

Perhaps the most sophisticated of all the rattler adaptations, and certainly the one that has earned its monster reputation, is the venom apparatus. The fangs work something like a hinged hypodermic needle, being attached to the head bone in such a way that they may be raised into a striking position or, when not in use, folded back against the roof of the mouth. The fangs are hollow, encasing a venom duct. They are an inch long and very sharp, but also very fragile and can be broken or damaged easily. Fangs are replaced frequently, every two or three weeks in some species. In the head of each rattler are at least half a dozen sets of fangs in various stages of development. Periodically the functional fangs are pushed out by the set growing in behind, much as the permanent teeth of a child dislodge the baby teeth.

Venom is produced and stored in glands located on either side of the snake's head. It is released when muscles squeeze the gland, forcing the venom down into the venom duct, and into whatever the snake has struck.

When it comes to lethal properties, real snakes can almost hold their own with mythical ones. Unit for unit, and depending upon the species, reptile poison can be 40 times as toxic as sodium cyanide, 30 times as toxic as typhoid endotoxin, seven times as toxic as the Amanita mushroom, five times as toxic as the venom of a black widow spider and twice as toxic as strychnine. Not only is it strong but it is buffered by some 25 enzymes, biological catalysts that speed up its effect and multiply problems arising from snakebite.

So complex is the venom-enzyme cocktail of the rattler, it has thus far defied precise chemical analysis. In consequence, venom is normally described in terms of the effects it produces. These vary but in a broad way can be summarized as follows. In a normal predatory situation, enzymes serve to predigest the snake's prey, as meat tenderizer softens a tough steak. Injected into a man, the enzymes begin to break down fibers and the cellular structure itself. They contribute to the hemorrhaging that often accompanies snakebite and are also responsible for "killing" the flesh in the area of the bite. Gangrene will sometimes be a by-product of this, causing chronic problems long after the effects of the venom have disappeared. The venom itself produces pain, swelling, nausea, allergic shock, hemorrhaging, weakening of pulse, lowering of blood pressure, increase in temperature, respiratory and circulatory difficulties and unconsciousness. Generally, if death occurs, it is caused by neurotoxic effects resulting in respiratory or cardiac failure or because hemorrhaging has so riddled the vascular system that the heart can no longer pump blood through it.

Klauber was of the opinion that the eastern diamondback was probably the most deadly rattler, not because its venom was the strongest (other species are more toxic) but because as the largest of the rattlers it could produce and inject more venom. With the possible exception of the pygmy species, under the right conditions any American rattler can inject enough venom to kill an adult.

All of which Americans had been aware of before they knew what an enzyme was. With lots of empiric evidence at hand demonstrating that snakebite was bad business, we have tried during the past four centuries an enormous variety of remedies—animal, vegetable and mineral—for snakebite. A list includes milk, eggs, tea, powdered crocodile teeth, horned-toad blood, onions, garlic, tobacco, indigo (and about a hundred other species of native plants), vinegar, turpentine, olive oil, kerosene, iodine, potassium permanganate, salt, gunpowder, ammonia, mud, opium, strychnine, ether, enemas, artificial respiration, song, dance, prayer and amputation. A remedy well thought of and still used in many parts of the country involves killing and splitting open a chicken or obtaining the heart or liver of a cow or deer. The bloody meat is then pressed over the wound. It is supposed to draw out the poison. It does not.

The most popular and persistently used remedy has been alcohol. Hundreds of snakebite victims have drunk themselves into stupors (and not infrequently drunk themselves to death). Klauber collected some of the whiskey dosages prescribed by physicians as well as folk healers. They include: two quarts of corn whiskey in 12 hours; seven quarts of brandy and whiskey in four days; a quart of brandy in the first hour, another quart within two hours; one-half pint of bourbon every five minutes until a quart was consumed; 104 ounces of applejack in four hours.

With this sort of treatment in general use, it is small wonder that a medical researcher estimated in 1919 that up to that time about 10% of the fatalities attributed to snakebite in this country were probably caused by alcohol poisoning. In fact, no species of booze is an antidote for snakebite.

The effectiveness of many of the bizarre remedies has been passionately defended, even by doctors. One reason stems from a peculiar aspect of rattlesnake behavior. The venom apparatus of the snake evolved as a method of quickly stunning or killing prey. As a predator, a snake is calculating, usually injecting only enough venom to get the job done. However, when the snake must defend itself it often panics. This is particularly true when it is striking at something as large as man, who is clearly beyond the potential prey range. Dr. Findlay Russell of the University of Southern California Medical School, perhaps the nation's leading authority on venom and snakebite, has suggested that in such a confused state the snake may respond erratically. It may eject venom in extraordinary quantities or, as in about one-third of the cases Russell was able to investigate, eject no venom at all. When this is true, when a venomless bite occurs, then any remedy—ice cream, spitting in the water or going to an X-rated movie—will appear to be an effective cure.

In the U.S. comparatively few people are bitten by rattlers each year—about 3,200—and very few, perhaps 11 or 12, die from the bite. Even so, the best advice is to avoid the snake. Even if one is intentionally looking for them, rattlers are not that easy to find. They are not abroad during the winter. In the hot weather they are most active after dark, when few people should be out in the bushes anyway. Heavy work boots, worn inside heavy pants, will turn most rattler strikes. One should be prudent about skipping lightly over logs, or putting a hand under a rock or on top of a ledge without first looking to see if it is occupied by a rattler. It is impolitic to molest, tease or play games with rattlers. This last bit of advice may seem superfluous but 25% of all bites occur not as a result of rattlers unexpectedly attacking men, but because men have been trying to handle rattlers.

In 1970-71 four of us conducted a year-long natural-history study project in the Huachuca Mountains. This area is especially good rattler habitat. We were in the field about 300 days, tramping through the scrub, in the canyons, in the flats. We encountered 32 rattlesnakes during that time. Only one, a grumpy black-tailed rattler that was disturbed while trying to catch a nap under a juniper, struck at us and he missed badly. But the prospect of rattlers kept us alert and added spice to the expedition.

The latter point was well made by a prospector named Van Horn who lived in those mountains for 45 years. Van was a hermit with anarchic tendencies. He was the natural-history guru of the mountains. He had also been bitten twice by small rattlers. "On neither occasion," he said, "did I like the results of the experiment. Both nailed me on the hand and both times I had a bad arm for a while. But it was my own damn fault. A rattler is like anybody else. He is just trying to get along as best he can. I will tell you one thing. As a pure device a rattler is a marvel. You can take a mouse or a little bird or even something big like a coyote or a bear. You can be interested in them and like them but you do not necessarily respect them. You may not like a rattler but you always have to respect him. It is a good thing to have something outside the human line to respect. It gives a man a sense of proportion."

Ghost Waters

For about forty years I have benefited from, admired, and therefore been a searcher after springs. In part this interest has been inspired by need. Both for recreational and vocational purposes, I have spent considerable time in bushy and undeveloped parts far beyond water mains and faucets. To do so for longer than a day, it is vitally important to come by potable water. Given the extent to which our ponds, lakes, and rivers, even the most isolated, are now laced with acid rainfall, radionuclides, leads, arsenics, PCBs, and other unpleasant to toxic pollutants, getting a good, safe, natural drink often requires finding a secluded niche from which flows sweet ground water. So motivated, I have hunted for and gratefully lapped spring water in deserts, mountains, swamps, jungles, woodlands, and prairies.

Even if thirst is not a consideration, springs are almost always nice places to visit. Some are prettier than others but I never recall meeting a truly ugly spring or one I didn't like. Springs are also invariably points of congregation for the cream of local societies of flora and fauna; for morning cardinal flowers, midday dragonflies, evening bears, and such. Finally, there are often instructive historical accumulations around springs: intersections of old trails and roads, bits of whittled and fitted rock, an occasional arrowhead or Indianhead penny that somebody lost while kneeling down to guzzle. Therefore they are apt to be good places for ghosts.

Before things get out of hand, a disclaimer should be entered. We are not talking here or subsequently about conventional—but I think fictitious—wraiths who supposedly amuse themselves perpetually with childish Halloween trips. I have never met such a being around a spring or elsewhere and am totally skeptical about reports of their existence, mainly because I think that the behavior attributed to these spooks is preposterous. It seems most unlikely that such obviously talented veterans would be content to spend eternity dressed in ragged mumus, living in damp basements, stuffy attics, and weedy graveyards; making their presence known by an occasional weak boo. Rather, with all their experience you would expect to find them, if they existed, in comfortable, probably luxurious homes, the better class resorts, or, if outdoorish, in scenic areas such as the south rim of the Grand Canyon. They would dress sensibly in easily laundered tunics or unisex body suits that did not catch on window latches during materializations. They would be fine conversationalists, probably and naturally most interested in eternal verity subjects but capable in lighter moments of being entertaining about such things as health food fads.

My theory about ghosts is that if you hang around things, places, events, and even ideas that have engaged our predecessors, you are sometimes able to get in touch with some of the thoughts, feelings, and concerns of the departed. Springs promote such associations because people have been hanging around many of them for long periods of time. Let us say you are looking for a rock of a very specific size and shape to chink up a hole in an old wall that encloses a spring drain. If you are in a mood and of a persuasion for soundless dialogues, you may get some mocking advice from a raspy, tobacco-hoarse voice that makes no sound waves. "Not that one, bub. I tried her in 1867. She didn't fit then and she ain't going to fit now."

More than fifty years ago a remarkable New York City resident by the name of James Reuel Smith summed up the case for springs, their appealing properties and the pleasures of hunting for them. He wrote: "Springs are attractive not only to the thirsty traveler, but also to the artist, the photographer, and the lover of pretty nooks and rustic scenery. In general the Spring seems to delight in picturesque surroundings, and its moisture freshens and encourages neighboring vegetation, and offers attractions that allure the denizens of the pasture whose presence redeems the solitude from loneliness without disturbing the restful stillness that soothes the admiring wayfarer. . . .

[A spring] is, for the meditative, a link which connects the thoughts with the past."

For me, a considerable bonus accruing from my sincere but sporadic interest in springs has been that because of it I have become acquainted with the works and (as above defined) spiritual remains of this James Reuel Smith. He was, so far as I am concerned, the most impassioned spring connoisseur and the champion spring hunter of all time.

Smith was born into an old New York family whose members, early in the nineteenth century, did so well in cotton, sugar, and banking that James Reuel (who was born in 1852) had, as family records delicately put it, sufficient "private, inherited means to retire early in life and devote himself to preferred pursuits." The most preferred of all Smith's pursuits was locating, contemplating, photographing, reading, and occasionally writing scholarly monographs about springs. So engaged, he spent many years traveling in Europe and the Mideast gathering information about the springs of Greece and Rome, of the Bible, and those of "romantical and poetical literature." However, his first and, I think, greatest aquiferous quest was a domestic one. For five or six years around the turn of this century he explored the northern districts of New York City, where he found some 150 springs. In 1916 he interrupted his classical works and organized his notes from this period into a manuscript and left a bequest in his will for its publication by the New York Historical Society. Smith died, full of years and presumably good water, in 1935; and in 1938 the Society issued his book, *The Springs and Wells of Manhattan and the Bronx, New York City at the End of the 19th Century*. Some twenty years later I came quite accidentally (so it seemed then, but now I am not so sure) on a copy which I have since read repeatedly for instruction, pleasure, and therapy.

As a true sample of the style and substance of *The Springs and Wells of Manhattan . . . ,* take the passage having to do with McCann's spring, which Smith first found on May 13, 1898. He noted that it was precisely 175 feet west of Broadway on 149th Street; that a copious flow of water drained from the spring into a "clean butter tub" that held fifteen inches of water. At the spring that day, wrote Smith, was "a large bent horned goat chained in a small outhouse, there are two geese very proud of a small yellow gosling and there are two dogs—one very big and ferocious, the other with a loud bark but a very gentle disposition. There are also some chickens and six children, nearly all the same size and about six years old.''

Finding and closely observing such springside scenes was the joy of Smith's life. However, he recognized that a complete description must include some social notes about how the spring was being used and by whom. Collecting such information must have been a torment, since there is much internal evidence in his journal that Smith was a shy man who found it very difficult to introduce himself to, much less strike up conversations with, strangers. A good many nineteenth-century New Yorkers who appear around his springs are introduced anonymously—''an old Italian woman,'' ''a Hibernian living in a shanty,'' ''a man with two cows who came through a gate as I was passing''—according to characteristics that could be gleaned without palaver by a keen but unobtrusive observer or an eavesdropper.

I have a vision of him, a neatly dressed little man, pacing nervously in front of a major spring site, trying to brace himself for an interview, his insatiable need to know warring with his shyness. Something like this may have happened around the above-mentioned McCann's spring which Smith visited several times. The land along 149th Street was being filled and graded to accommodate a block of apartments, and Smith was curious about what had happened to the water from the clean butter tub which previously had flowed in a little stream down into the Hudson River. He never, according to his notes, found courage to ask McCann, the man for whom the spring was named, but one day did get into conversation with a gentleman identified only as ''McCann's Irish father-in-law,'' who told him that a line of wooden barrels had been laid under the fill to carry away the water. ''McCann's Irish father-in-law says they will have trouble with those barrel sewers as sure as the Divil is in Hell.''

Again, like a few old flickering movie frames, there is an image of James Reuel Smith politely tipping his hat to the profane Hibernian and slipping off to add this note to the store of information he was accumulating about the spring.

One day I was fortuitously able to get some first-hand confirmation of the impressions of Smith which I had formed from reading his accounts of spring hunting. During his later years, he and his wife lived in a rambling Victorian house on a bluff in Yonkers overlooking the Hudson. The house was named Dondona, after a sacred spring in western Greece. Out of curiosity about Smith, I made a special trip to the neighborhood. Dondona still stands and is being carefully restored by a young couple named Loma who were unaware that it was once the property of the world's greatest spring hunter. However, just down the hill lives, as she has for the better part of the last half century, Mabel Lower, who has a lot of sharp memories of her former neighbors, the Smiths.

"He was a tiny little man with very short arms. No matter what time of day he always dressed like a banker. If you met him he would raise his hat and speak politely but he would scurry back to the house. When he was here he spent all his time with his papers and books. Delivery boys would come several times a week with new books for his approval. He did not like going into stores. Also he did not like automobiles, movies, or the radio. His wife would come down to my house to listen to the radio and sometimes dance all by herself to the music. She was such a cute little thing, but lonely I think. They had very few callers, and when somebody came to the door it was she who always met them. He would stay hidden in his study, but when the visitors left, she said, he would come out to ask lots of questions about them. If she was feeling snippy she would not satisfy his curiosity and tell him that he should have met the people and asked his own questions. I think he was a goodhearted little man and certainly a curious one, but he was so shy."

Reading and reading between the lines of old books is one way of picking up some information about the departed, and talking to someone like Mrs. Lower is a better one. However, about the time I finished reading *Springs and Wells of Manhattan* . . . for the first time, I decided that one day, both for the pleasure of it and as a means of becoming truly acquainted with Smith, I was going to have to hunt for at least some of the same springs he did and, so far as our disparate circumstances permitted, do it in the company of this shy, brave, gentle, independent-minded little man.

The expedition was long delayed, in part, truthfully, because of timidity. For reasons of temperament and experience that do not signify here, it has always seemed a fairly simple matter to get to, say, Lake Conowotyo. (You just paddle up the Yellowknife River for a week, turn right at the first big caribou herd, and follow a wolf trail across the Arctic Barrenlands.) However, I have always been admiring of people who casually report that they have made a roundtrip to West 149th Street. For anyone who lives at Iron Springs, Pennsylvania, the idea of wandering about New York out of sight of Radio City, finding obscure trails, coping with exotic folk and folkways, is scary. Getting up for it required a lot of self-goading. Also, there was a faint gentle voice that provided occasional encouragement. "Come now, Mr. Gilbert, buck up. Bold is the spring hunter. Have I related to you my encounter with McCann's Irish father-in-law?"

In any event, there came a time when no more excuses would do. Bidding a fond farewell to the comforts and conveniences of Iron Springs, stuffing my pockets with long, green, negotiable trinkets for the natives, clutching the dog-eared copy of *Springs and Wells* . . . , I set off to see how or if the waters still ran in the city wilds.

In the seventeenth century, there was probably no narrow, fourteen-mile-long island anywhere that was better endowed with fresh water than Manhattan. It was covered with loam, layers of sand and gravel laid down by glaciers. Lots of rainwater could be absorbed by these layers and would percolate easily through them. Beneath, there were gneiss and schist foundations, the roots of an ancient mountain, filled with cracks and crevices that acted as natural reservoirs. Writing after his exploratory visit in 1614, Adrien Van der Donck noted that "water flows out of the fissures and pours down the cliffs and precipices," and was "pleasant and proper for man and beast to drink, as well as agreeable to behold."

James Reuel Smith estimated that at the time of the first European settlements there were three-hundred full-fledged springs in Manhattan. However, it was a fragile system. Surrounded by salt and brackish water, the only source of fresh supplies was the rain, and subsequent building and paving cut down the area that could absorb rainfall. Excavations and tunneling into the rock foundations breached and diverted the old underground courses. Finally humans began using water faster than it accumulated in the ground. (By 1750, Manhattan was suffering from water shortages and they have continued sporadically ever since.)

"During the last decade," Smith wrote sorrowfully of the 1890s, when the population of the city increased by half a million, "springs and other natural features of the landscape are disappearing from sight with such celerity that it is merely a matter of months when there will be none whatever left in view upon Manhattan Island." Events have proved that Smith was overly pessimistic, but understandably he had a sense of urgency as he went about his self-delegated role of cataloging the last of Manhattan's free waters.

Smith did not spend much time exploring the lower part of the island—below 70th Street—because it had already become so built up as to be a wasteland from the standpoint of a spring hunter. He knew of some wild waters running in downtown tunnels and basements but did not think it appropriate to "describe or even enumerate the numerous springs so imprisoned." He searched only for open springs, "out of doors, visible and freely accessible." Since it had been good enough for the master, I followed the same rule during my own expedition.

Spring hunting may well have contributed to Smith's retiring nature, for unquestionably there are shy-making aspects to this sport. Take, for example, the situation on East 79th Street. In March 1898 Smith found a spring from which "water boils up lustily" at the edge of the sidewalk on the north side of the street, 150 feet west of East End Avenue, adjacent to a blacksmith shop. This being the most southerly open water Smith located on the east side of the island, it seemed like a good place to try to pick up his trail. Not unexpectedly the blacksmith shop is gone, but like some portentous omen, exactly 150 feet up 79th Street, a thin stream of water trickled across the sidewalk. It seemed to issue from a five-story building, the insides of which were being vigorously gutted by demolitionists preparatory to something else being built inside the shell.

"Hey, you there. Yeah you. Whaddaya want?"

"I was wondering about that water. Where it comes from."

"Where ya think it comes from—the roof? It comes outta da ground in the basement. Like a river. You from da Depatmnt or what?"

"No. No. What I am . . . I mean I found this old book and it says there used to be a spring just about here and I was. . . ."

"Yeah—sure. LOOK OUT!"

"What?"

"Look out for your head. You get it smashed in. You can't stand here. This is hard hat. LOOK OUT."

This is the kind of intimidating thing that a rookie urban springologist expects will happen everywhere, but in fact the encounter was the only even mildly hostile one to occur in several weeks of spring hunting. Despite their ferocious reputation, the natives seem more inclined to be curious than suspicious about visitors from the larger world. Far more typical of the islanders than those of the 79th Street band were two named Bill Lord and Wade Pendleton. They were met early one morning just inside Central Park, opposite 104th Street at about the place where on October 26, 1897, James Reuel Smith located a small spring that fed into a stone horse trough.

When first sighted, Pendleton was riding on the shoulders of Lord, who was running laps around a small, bone-dry ornamental pool in the middle of which stands the battered bronze statue of a nymphlike figure.

"Good morning, my man," Lord puffed politely and then put his companion down and sat with him on the cement bank of the empty pond.

"What are you guys doing? Getting ready for a horse and rider contest?"

"I am the trainer," explains Lord, "and this is my runner," introducing Pendleton. "He has just run as far as five miles in the park and is tired. I was letting him cool down easy, giving him a ride around this old lake."

"You're joggers?"

"Man, no, no. I am your jock," says Pendleton, making an appropriate gesture. "This summer when I am trained I will run for El Barrio."

"What distance?"

"Maybe short, maybe long, the half mile. Whatever is best, man. What are you, man?" Pendleton inquires hospitably.

"That's complicated. What I'm doing is I've got this old book. It says that someplace right around here there was water coming out of the ground, a spring. I'm trying to find the place. See, here it is. The old guy took a picture of it."

"Man that is cool—those old pictures, but it don't look the same, the plants are different."

"They'd change but the rocks seem about right."

"It says," says Lord skimming Smith's account, "104th Street. It was probably over there, way across 5th Avenue."

"But it says it was in the park."

"Man, you don't understand this place. Once this was all park. First dude to see it was Henry Hudson. When he came up here everything was just a nice park."

"I still think that spring was right about here. Maybe where that pond and fountain are now."

"I will not hassle you, man. That may be true. Somebody could mess up the water like they do everything. Look at her," says Lord by way of illustration, pointing to the remains of the bronze nymph. "She had a flute once. They rip that off her. As you see, they bust off her arm, too. What I'm going to do is give that lady my shirt so she won't think nobody care about her no more."

Lord pulled of his sweaty shirt and hung it artistically around the shoulders of the armless, fluteless nymph. "Now you, my man, have a very good day." He gave the customary native farewell and then jogged off into the park with his runner.

The most southerly open spring Smith found in the 1890s on Manhattan was also in Central Park, 160 feet west of East Drive, opposite 76th Street. Satisfyingly it is still here and surprisingly it seems very little changed from the day Smith photographed the place. It still rises out of a grassy bank below the Ramble and northeast of the Central Park lake. It fills a dishpan-sized, unwalled, apparently untended hole in the ground. The overflow spills slowly out of the hole and trickles downhill, leaving fifteen feet or so of soggy turf. Then the water is reabsorbed into the ground. Smith wrote that it was a "quiet" spring that "fits so snugly in its grassy slope that the only likelihood of discovery is from chance reflection of the sun from its surface."

Given Smith's explicit directions, the main difficulty in getting to this water now is edging through the throngs of people who are milling around the area on a nice Saturday morning. In the vicinity there is a demonstration-gathering complete with TV and police crews, a bicycle race with marshals and spectators, and of course the joggers who are everywhere from Sputyon Duyvil to the Battery, from the Hudson to the Harlem. There are lean joggers and stout ones, graceful and awkward, he and she, old and young joggers, joggers in Olympic-style costumes and joggers in house dresses and fedoras. There is a mated pair who have put their toddler in a park sandbox and then, rather than sit supinely watching him play, are getting in their jogging—around the sandbox at the rate of approximately fifty-three laps to the mile. Unbriefed, a stranger might think that the city was in the grips of some religious mania such as set medieval communities to shaking and twitching. Here it is the up and down bounce, the pumping arms and lolling head.

Two joggers, taking a blow before trotting up the slope into the Ramble, are the first to find a spring hunter squatting over, measuring the little pool in the grass. (The water stands five inches deep.)

"We going to get that broken sewer fixed? Somebody is going to slip in the mud and pull a hamstring."

"That's no sewer. That's a spring—part of your natural heritage."

"You mean a real spring? Like in the Poconos or Aspen or someplace?"

"A genuine spring. Look, here's a picture of it in this book like it was eighty years ago."

"Hey, Michelle. Look at this. This is real water. I always thought it was a sewer."

"That's lovely. New York just turns me on. It's so, like, historic."

"I mean really. Thanks for telling us. It's like a kind of rush to start the weekend. You have a real, but real, good day."

There is at least one other spring in Central Park, now such a remarkable one that it seems prudent not to be too specific about its location. Hunting for and especially finding it will be rewarding for beginners, and certainly any devotees of James Reuel Smith will already have found it. Smith and many other of his contemporaries knew it as Tanner's Spring. It was named after a Dr. Henry S. Tanner who in 1880, for purposes that are now obscure, gave a public exhibition of fasting. Tanner, it is said, went forty days without eating solid food but drank daily from this spring. "Tanner's apparent ability to live without eating was attributed to some nourishing elements in this spring," wrote a skeptical-sounding Smith. "At any rate, people came even from distant parts of town with bottles, pitchers, and pails, which they filled here and carried to their homes."

Later the Health Department tried to bury the spring, but Smith said it had "a rebellious tendency" and kept popping up on the surface. It still does and is still flowing, welling up from under one boulder, flowing across another and then going back underground in a woodsy, very out-of-the-way corner of the park. In addition to the historic waters, this glade is extraordinary because of its neatness. There is not so much as a flip-top can ring, a sandwich wrapper, or even a fallen branch visible. The place looks as if it had been manicured in preparation for the filming of a TV nature special. No camera crews turn up, but the manicurist does. She is carrying a rake and a shopping bag, in which there is a little junk, a few twigs, and some pebbles that strike her as unsightly and therefore are being removed. She is Bessie and asks, "How you like our place?"

"I like it fine. Whose place is it?"

"It belongs to my boss. I do the work."

"Who is your boss?"

"Honey, who else? The Lord God Almighty."

"Oh."

"More years ago than I like to think about, honey." Bessie Nelson came to the city from South Carolina. She worked for some time as a domestic, but nine years ago the Lord directed her to this acre of park. ("It was bad then, full of weeds and trash and dog stuff, number two.") The Boss told her to clean it up and keep it clean. Ever since, three or four hours a day, seven days a week, 365 days a year, she has been so engaged. She cares not only for the flora but the fauna in God's Acre, having named the local pigeons, jays, and squirrels. She has become remarkably intimate with them, to the extent that when one is called—"Your turn, Thomas"—a creature will light on her outstretched hand or crawl up on her shoulder to be fed snacks.

At first, Bessie says, park employees were skeptical about her activities, but they have become cooperative, loaning her tools and giving a hand with some of the heavy tree-trimming work. "One day the big man, in charge of all of this, he come by. He says to me, 'Bessie, you a worker. Why don't you work for us?' I say, 'Oh no, indeed. I can't work for no more than one boss at a time and I got my Boss a long time ago.'

"Bessie, in this book it says that an old-time doctor thought if you drank the water here you could go for days without eating."

"I heard something about that."

"How did you hear about the old Doctor? His name was Tanner."

"Indian man tell me."

"Who?"

"Just an Indian man who use to come here. We make a lot of friends. People used to be here and they go away. People from Colorado and Australia, they come back to New York and they come right out to Our Place. It's good here."

"I'm from Pennsylvania. I know I'm going to come back again."

"I do so hope. You have a good day now and the Lord bless."

Deep under the pavements, buildings, and rubble piles some of the old waters, which once broke out on the surface as springs, may still flow. However, their courses have been so buried and altered that now only deep constructionists and demolitionists can find or are concerned with them. A spring person following after Smith and following his rule—that is, counting only open, free water—is confined to searching in small, never or lightly developed plots that can still absorb and discharge water. Fortunately, there are more of these surviving than Smith gloomily predicted there would be, more than a modern pilgrim expects there to be. (As a matter of record, sixty-two of the Manhattan springs sites that Smith knew were investigated, and including two probables—patches of damp earth—there was some evidence of water at nine of the original locations.)

One of the best hunting places is St. Nicholas Park, more or less in the middle of Harlem. In comparison with the famous and chic Central Park or the elegantly designed and maintained public grounds around Fort Washington, St. Nicholas has a shaggy, unkempt appearance. Domestic turf has been overwhelmed by wild weeds: docks, plantains and onions. There are considerable stands of old, unpruned maples, locusts, oaks, and beeches, whose big boles are scarred but not weakened by the initials and tender sentiments carved into them. Barberry and privet once planted in ornamental hedges have sprawled and spilled between the trees, making jungly thickets that extend onto a steep rocky cliff that separates the western edge of the park from the campus of the City College. Although half a million people live within a few blocks of it, there is a curious wilderness quality to St. Nicholas Park. It gives a sense of being its own, rather than municipal, ground, of maintaining itself through its own coarse, thorny devices.

Many of the originally planned walks are in great disrepair, having been buckled by frost and severed by erosion. However, new paths have been cut by people who want to get into the woody meadows, back into secret hideouts to find quiet reading, thinking, or trysting spots. According to Smith, the children of the 1890s, playing at scout, explorer, or maybe Rough Rider, had already made a maze of trails along the St. Nicholas cliffs. In this "brambly environment" he found several springs in 1898 and '99. At least one still flows weakly. "The children," wrote Smith, "say they used to drink out of it until other children made it muddy."

"You guys ever drink that water?"

"Hey man, you crazy? That is bad. That is poison."

There is more water farther down the slope, apparently where Smith found another spring. However, its situation is much different and it takes some finding, in fact requires a little urban spelunking. At the spot Smith describes, there are the remains of a cistern—a raised manhole-like masonry structure. The cap has collapsed and the hole stands open, but there are still some steel spikes implanted in the rockwork. They lead down into a cavern littered with chunks of broken cement but otherwise open and clean. The rocks are damp and get wetter the farther the tunnel is followed back into the hill. From the darkness ahead, beyond the point where it seems prudent for an unprepared caver to proceed with only a packet of damp matches for light, there comes the tinkling sound of water dripping into a small pool.

Johnson and his dog Sam, a big shepherd type, are waiting curiously on the surface when the spring hunter emerges. The mission is explained and Johnson looks through the passages in the old book which have to do with St. Nicholas Park. "I'm not surprised," he nods. "This is a historic place. Did you know that the Vanderbilts and those other old big shots used to race horses right down there on St. Nicholas Avenue?"

"No, I didn't."

"Yes indeed. With your interests, you might like to walk with us to the other end of the park. We go down there to meet a friend of mine, Clarence, who has a Rottweiler, a lady dog, who is a friend of Sam's. We meet at a place I think you will enjoy."

Johnson says he has lived in Harlem, between 128th and 145th Streets—5th and Lenox Avenues—for the past forty years, and for the last ten of them he has been coming each morning and evening to the park to walk with Sam. "He knows when the time arrives. It would break his heart if we missed our walk and because of him my heart, too."

"Mr. Johnson, you hear a lot of bad comments about city dogs."

"I can appreciate them. Many people just like the idea of keeping a dog, but not the responsibility of keeping them. The dogs become trouble for them and for other people. You understand me. In my case it is all a pleasure. Put it this way. I would have been a poor man these last ten years without Sam."

Johnson says there is a plant in the park that sends up shoots in the spring. Sam eats them as an aid, Johnson thinks, for digestion. "Sometimes you see Chinese fellows collecting those same stalks." By and by one of these plants occurs. It turns out to be poke, the preeminent wild pot herb.

"Now that," says Johnson, "I am pleased to know the name of that plant. Let me ask you something else. You say you are a country man. Several times just at dark I have heard what I thought was an owl. Is that possible, in your opinion?"

"I'm no authority on bird life around here, but I don't see why there couldn't be an owl. There should be starlings, sparrows, and some rodents around for prey."

"We do have some rodents in these parts."

Shortly Sam spots his friend the Rottweiler sitting by the side of Clarence, Johnson's friend. The two dogs nuzzle and sniff affectionately and then trot off to drink from and splash in a perpetual puddle fed by a thin streamlet that rises at the base of a thicket, just about where Smith found a spring in the fall of 1899.

"You see," Johnson says, grinning, "why I thought you'd like this place?"

This is the wildest-looking and -feeling spring in Manhattan. If the sounds of the city that swirl all around the narrow frontier of the park are ignored, it might be one rising in a high Sierra meadow or from an Appalachian clearing. Some logs have been dragged up to it for seats. In front of them are the remains of a fire and some peeled, blackened sticks used to toast hot dogs or marshmallows. Part of the fire was made from a one-by-six plank. Enough of it is unburned that the stenciled words "Do not . . . " are still legible. Probably the camp was illegal in some fashion, but it is hard to get worked up about the criminality of sitting around this spring and a fire, maybe listening to an owl. St. Nicholas Park seems to be a place where the natural law has superseded municipal ordinance.

For obvious social and hydrological reasons, most of the remaining springs, seeps, dribbles of wild water in Manhattan are to be found in parks, but not all of them. For example, Smith found a number of springs on the southern and eastern flanks of Harlem Heights, along what is now 110th Street, the Roosevelt and Harlem River Drives, the approaches to the George Washington Bridge. None now seem to remain in the original locations but there is still free water on the faces of the artificial cliffs created by the road cuts. It oozes over the rock, like coffee out of a cracked cup, giving evidence that some of the old aquifers still persist in the foundations of the island.

One of the most watery non-park areas is along the last two blocks of Dyckman Street, just before this narrow, brick-cobbled thoroughfare dips down and ends at the Hudson River. At the head of this cul de sac, where Payson Avenue intersects Dyckman, there once flowed Whitestone Spring, which, according to Smith, was "demonstrated by many analyses to be the purest on Manhattan."

If they still run, the pure waters of Whitestone now can only wash the deep foundations of a large apartment complex that rises from the former spring site. Nevertheless, water still weeps out of the southern face of Inwood Hill, which Dyckman Street hugs. In consequence, the slope and rocky outcroppings along the sidewalk are covered with mosses, some ferns, and even a clump of saxifrage. It could easily be a bank above a trout stream in the Catskills.

James Smith came down this wet walk in 1898, looking, of course, for a spring, one that he had been told rose at the very foot of Dyckman in a sandy cove on the shore of the Hudson. "The little Sandy Bay of 1614 is there still," wrote Smith after he found it, "remaining placidly to witness the truth of those [Adrien Van der Donck] who wrote of it nearly three hundred years in the past."

Eighty years later the beach is covered first and more or less permanently with a layer of oil and mixed garbage that has washed in from the Hudson and down Dyckman Street. On top of it, on this particular day, there are also the carcasses of eleven waterlogged, very dead chickens, origin unknown. Even so, it is possible if not pleasant to scrape aside this mulch and poke down to the sands of 1614 and 1898. In them, at the top of the unsightly beach, there is a trickle of water (and oil), the delta, so to speak, of the seep springs that ooze down the hill along Dyckman Street. Aesthetically and probably by analysis, if such were ever made, this is the worst bit of open water left in Manhattan or maybe anyplace. Still, there is something appealing about it. The spring at "Sandy Bay" must be one of the spunkiest and most tenacious on the continent.

In 1897, according to Smith, the largest spring in the city was located at the extreme northern end of Inwood Hill. Known as Cold Spring, it rose at the tip of Manhattan Island about a quarter of a mile east of where the Henry Hudson Bridge pilings now stand. It flowed out of a fissure of rock on the hillside and drained a hundred yards or so into Spuyten Duyvil Creek, which had already been enlarged into the Harlem Ship Channel. Cold Spring was walled in and filled a six-by-three-foot stone basin. Its proprietor was a cantankerous old man, by all reports, named Andrew Seely, who lived there in a cottage and made a living selling bottled drinks to boatmen using the river and ship channel. Smith judged that there was not another dwelling within a mile of Seely's, and though the city has moved somewhat closer, this is still perhaps the most isolated spot on Manhattan.

Not even the foundations of Seely's cottage remain, but where it stood can be guessed at with fair accuracy. There is a small triangle of flat, wedged in against the hill and the creek, which would be the first available building site inland from the Hudson. Since Seely's time a grove of sizable oaks, locust, and willows have grown up on this site and they screen it from the soccer, baseball, kite-flying, dog-walking grounds of upper Inwood Hill Park, a large recreational prairie that stretches eastward for a half mile toward Broadway.

The spring at Seely's old place is gone, but that comes as no surprise since it was going when Smith last visited in June 1898. By then Seely had buried the drain and closed over the spring so that no one could use the free cold water as a substitute for the bottled drinks he was pushing. Later some authority apparently attempted to complete the closing of Cold Spring. Portions of old conduits are visible. Probably they were laid to carry the spring water down into Spuyten Duyvil, and discharge it below the creek level. However, flowing water is a powerful and persistent thing. Several hundred feet inland from the flats but in the same drainage system, on the north side of Inwood Hill, more water has broken out of the ground. This spring is not nearly so large or grand as the old Cold Spring was, but it forms enough mushy ground and a sizable enough puddle to attract a dog and two small children; also to exasperate the parents of the children and the owners of the dog. Alternately they command and warn. "Get out. That water is dirty. Get out of that dirty water."

If anything, this new (if it is) spring is located more historically and in a better place for ghosts than even Cold Spring was. It rises in what seems would have been the center of Shorakkopoch, which was once the largest Indian village on Manhattan. Legend has it that here in 1626 Peter Minuet negotiated to buy the whole wet little island for sixty guilders' worth of trinkets.

Presumably this spring was not visible in Smith's day, because he did not describe it, but he would have liked it and maybe enjoys it now. In connection with his comment that springs move the meditative to think of the past, he wrote, "Anyone may today sit down at the brink of active bubbling springs on Manhattan, and read references to them in the journals of travelers, in old Indian deeds and in the patents of early Dutch governors."

There is a final, very obvious attraction to springs. Once having found one, there is a great urge to drink from it. Again Smith explained it best: "Greatly satisfying indeed is the draught from a spring where none is said to exist, and which has been come upon after patiently and inductively following a trail marked only by a moistened stone here, a willow farther on, and then a piece of watercress. In the days, not so very long ago, when nearly all the railroad mileage of the metropolis was to be found on the lower half of the Island, nothing was more cheering to the thirsty city tourist afoot or awheel than to discover a natural spring of clear cold water, and nothing quite so refreshing as a draught of it."

Smith himself sampled and wrote, like a serious wine taster, about the vintage Manhattan spring waters of the 1890s — "cold and pleasant to the taste" (115th Street and Lenox Avenue); "white in color and rarely freezes" (138th and Lenox); "sparkling . . . like champagne bubbles" (West 149th Street at the Hudson); "peculiarly tasteless if not actually flat. It has no frogs" (Whitlock Avenue); "bright sparkle but the water is comparatively warm . . . however when Commodore Vanderbilt was ill, the water from this spring was drawn fresh twice a day and sent down to him, his physicians have declared it to be the best water to be found between New York and Tarrytown" (Fort Tyron Park).

Even in Smith's day, sampling open waters in Manhattan was risky by reason of contamination and in most places technically a misdemeanor according to the municipal water regulations. In the past eighty years the situation has worsened considerably. Officially there is no open water left which is safe or legal to drink. Environmental conditions generally make the reasons for this prohibition obvious. Only the most blatantly suicidal would, for example, think seriously about squatting down on the "sand" beach at the foot of Dyckman Street and swigging from the water there. However, the idea, the challenge of such an experiment keeps bubbling in the mind of anyone following after and perhaps with the master. Finally there comes a place and time when it can no longer be repressed.

Put it this way — at a certain place on Manhattan Island (to be no further specified so as to avoid problems of law, public health, and conscience) there is a bit of flowing groundwater from which a pilgrim can drink as James Reuel Smith did eighty years ago in the same spot. (Currently the water is clear but bland, enlivened only by a faint back taste of carbon monoxide.) At least one did so and some months later remains in good health despite the act, and spiritually much exhilarated because of it.

Cold

Potter is a sparsely settled county in north central Pennsylvania. It sits astride the crest of the central Appalachian highlands. Its springs, icy brooks and white-water streams flow through the Ohio River system to the Gulf of Mexico and through the Susquehanna to the Atlantic. The upper ridges down which the waters spill are only 2,500 feet or so above sea level, but the mountains of Potter are formidable. They are laid out in bewildering complexes of intersection ridges, sharp valleys, gulches and ravines; littered with rock slides and boulder fields; chopped off into cliff faces and cut by streams, sinks and bogs. This topographic maze is overlaid and further complicated by an impressive northern hardwood forest. There are great 80- to 100-foot black cherry trees and mixed with the cherries are maple, oak, ash, birch, beech, poplar, pine and, in the bogs, tamarack. In the understory are thickets of laurel, button bush, ferns and all manner of lesser vines, flowers and brush.

At the center of Potter is a quarter-of-a-million-acre tract officially designated and supervised as the Susquehannock (after the Indian tribe of the same name) State Forest. Locally it is often called the Black Forest. The name is appropriate. There are ravines and glades within Potter's big woods which, because of the extensive and deep canopy of foliage, see very little of the sun except the pale cold, winter one.

The Black Forest is the domicile of a good many bears, bobcats, foxes, mink, raccoons, porcupines, beaver, ravens, goshawks, grouse, turkey, timber rattlesnakes, trout and a lot of deer—a herd of some 35,000. In contrast, there are not many people in the 698,000-acre county—about 16,000 residents, which works out to almost 39 acres of woodland per person.

As might be expected from what does and does not live there, the weather in Potter is also wild. A set of meteorological maps, which illustrate such things as temperature and the amount of solar radiation and precipitation, will show that Potter has a meteorological profile similar to that of northern Maine, Minnesota, central Alaska and a lot of Canada. From December into March the average temperature is well below freezing, and throughout the year there are fewer than 150 dependably freeze-free days. Sixty to 100 inches of snow falls each year, and it is not unusual to find snow around the ravines in May. Despite the fact that it lies at about the same latitude as Chicago, Des Moines and Providence, R.I., the climate of Potter is what might be called subarctic. In part, the elevation accounts for this anomaly; also the fact that these highlands are the closest ones to the eastern Great Lakes, across which arctic winds and air masses proceed unimpeded until they hit the mountains.

Especially in Pennsylvania, but also in adjacent parts of Ohio, Maryland, New Jersey and New York, people will sometimes say that they are going to be, or have been, "up in Potter," (the word "county" is almost never added) or "That buck looks like a fawn alongside what they got up in Potter," or "By God, I bet you'd freeze your patootie this morning if you was up in Potter."

Close examination may show that they actually have been up in Clinton or Cameron, which are adjacent and very similar highland counties. However, Potter has such a reputation for cold, solitude and wilderness that it has become a kind of code word. It also has come to have a certain mystic meaning. A fellow who says he is going up to Potter is hinting that not only is he traveling through space to an exceptionally wide, cold place, but also backward through time toward a point when everything, or so we like to imagine, was more elemental, somehow purer, than it is now.

Potato City

U.S. Highway 6, running east and west, more or less bisects Potter. Halfway through the county, set back from the highway so that the Black Forest is at its hind end, is the Potato City Motor Inn. It provides the grandest accommodations in all Potter; in fact, accommodations that would be regarded as better than adequate almost anyplace. In addition to the Inn there is a Potato City Trap Range, a Potato City driving range and a Potato City Airport, but there is no Potato City. The reasons for all of this are curious.

The bulk of the cleared agricultural land in Potter lies to the north of Highway 6. The principal commercial crop is potatoes. Shortly after World War II some big potato men found themselves embarrassed because when potato magnates from other parts of the state, or from Maine or Idaho, came to visit, there was no suitable place for them to stay and be entertained. So the growers and packers chipped in and built the Potato City Motor Inn and associated facilities. Their idea was that it would make them a nice club and would also pay for itself by taking in non-potato travelers. Clubwise it was successful, by report, but it didn't do well with travelers. There were not many of them, and the potato group didn't treat them with great hospitality when they did show up. By and by, the potato men sold the property. It has since passed through several hands and has ended up in the capable ones of two young couples from southeastern Pennsylvania who have considerably refurbished and expanded the business.

The social heart of the Potato City Motor Inn, one that hints at its original function, is a convention-sized eating, drinking, dancing and general-carrying-on complex. Despite its grand proportions, this is a fairly chummy hall, largely because of a fine, large horseshoe-shaped bar and a large flagstone fireplace. A massive young friend named Sam and I were sitting at a fireside table there on the last Friday night of January.

This was almost everywhere a bitter and extraordinary winter. However, it was less remarkable in Potter than elsewhere because it was more common. That particular day it had snowed six inches, the thermometer had not risen 5° above zero and the wind had been blowing briskly off the Great Lakes. It was, in short, an average Potter

January and nobody at Potato City was paying much mind to the weather. Sam and I certainly were not, being fully and pleasantly occupied with a couple of strip sirloins, hot potato salad (anything at Potato City that has potatoes in it is likely to be very tasty), Genesee beer, and later on coffee and sour-cream raisin pie.

The only other people there were a dozen or so snowmobilers, a good many of whom truck on up to Potter on winter weekends from all over the Mideast to compete on the racecourse laid out behind Potato City or to cruise the trails the state has laid out and maintains for them in the Black Forest.

Generalities are suspect, including this one, but generally snowmobilers seem to be hearty, down-to-earth, convivial folks who are fond of pickups, campers, 4WDs, outboard motors and CB radios as well as their snow sleds. They tend to run in clans or clubs with names like the Cool Drifters or the Blizzard Beaters. These names and appropriate logos are usually stenciled on windbreakers, helmets and vehicles that are color coordinated in Day-Glo purples, oranges and yellows.

What with some back-and-forth between tables, questions arose in an amicable way about what Sam and I were, if we weren't snowmobilers. We said we were not anything identifiable in an organized sporting way, but that what we were doing was stoking up on the victuals preparatory to setting off in the morning to snowshoe through the Black Forest.

"You're not going to get far falling around on those things," one of the snowmobilers said. "Youll just get out and have to turn around and come back to get in before dark."

We said we weren't coming in at least for a few days.

"You mean you are going to stay out *nights*?"

The admission that that was what we intended brought down the house. Nobody around Potato City had ever considered the possibility or met anyone who wanted to go off and bed down in the woods in January.

"My God, you'll freeze to death."

"Maybe, but that's not the plan. We've done it in colder places."

Inevitably the question of why we were doing it came up and we answered that this was our idea of having some fun. Whether or not they have ever cuddled up in a drift, people know this is a lie. There is no way such activity can be fun in the commonly accepted meaning of the word—not fun like sinking a long putt, whizzing along on a sled or eating good sour-cream raisin pie.

At least part of the whole truth might have been offensive in our particular circumstances. One reason, the least attractive, for wandering about the subarctic mountains in the winter is showing off to people who do not, a group that includes most of the sane citizenry. Admittedly our cool woodsman image gave us some pleasure and impressed the snowmobilers, who until that moment considered themselves the ultimate winter masters. But if you go in much for that kind of bragging, you are always vulnerable. For example, anybody who walked into Potato City that night and started talking about a four-bivouac climb on McKinley, the last one of which was spent hanging by one piton and a chock made out of a freeze-dried pork chop, could have cut some big notches in the two of us.

There were better reasons, or so we hoped, for doing what we were going to do. When everything has been said about how sport, games and recreation build sound bodies and promote good citizenship and television careers, the fact remains that this kind of arbitrary, practically unnecessary activity has been popular with people for as long as we have been people, because it provides escape and therapy. For reasons of no general interest, Sam and I both were in need of escape and therapy, and for sundry reasons of temperament and experience, cavorting about in the winter wilds of Potter seemed like a quick way to get some.

Sam

For the past 35 years or so I have been fortunate in having a fair number of friends and acquaintances who have shared my needs and opinions about recreation, escape and therapy. With the passage of time, however, a lot of my old cronies have been struck down with bad backs, caught up in soft sheets and become too dead for this sort of thing. The circle with whom I can, or would want to, make winter snowshoe trips has dwindled down to four. At this time, one of these is doing something important for the governor of Alaska, another is an art student in San Francisco and still another is in Vermont looking for work in maple sugar. The fourth and only practical possibility is Sam. He lives nearby and is an orchardist who throughout the hard winter has been pruning apple trees from the bucket of a hydroladder.

You can play tennis or catch with people who are more or less all right, but when it comes to high-class freezing and fatigue, you have to have a partner who is absolutely suitable. Mostly, it is a matter of being and knowing that you are peers. For example, it will not do if only one can start a fire in a blizzard or properly set up a tent in a drift. The one who cannot will feel patronized and the one who can, unfairly burdened. Either way, tension and trouble will develop, and there is usually not enough energy, physical or psychic, to spare for this sort of thing. Especially, you have to have very compatible opinions about what is interesting, exciting and funny. Otherwise, in inevitable moments of misery you will turn on each other and simultaneously say, "It's your fault."

Beyond being very suitable in these respects, Sam is a notable horse, being 25, 6'4'' and 230 pounds. Except for an occasional small NBA forward or an especially agile tight end, Sam is physically the most impressive man I know. He may be even more impressive than these professional athletes because he works harder than they do and has not been hurt so much. He can lift a garden tractor out of a hole or unload a hay wagon two bales at a time. Anyone who does not find such things impressive should try them a time or two.

Even at 25, when I was more of a horse, I was not such a one as Sam is, and now at twice his age I may not be half the horse he is, but there are different elements of peerdom. For example, we move along on snowshoes at about the same rate because I have been doing it for a long time. Sam can wrench a four-inch dead tree off its remaining roots, but we get a fire started in about the same time. Out of necessity I have given up wrenching and developed a pretty good eye for squaw wood, which can be broken over the knee. Sam meets the cold head on—toughs it out. I do tricks in my head to finesse it. We have different edges but the sum of them is quite equal, a fact we recognize and find appropriate but not a matter for competition or envy.

Dressing Up and Down

The morning, the next to the last one of January, was more or less what the previous night had promised it would be. There were drifts halfway up the first-story windows of the Potato City Motor Inn. The temperature stood at $-14°$ and, what with the wind, the chill factor was about $-50°$. As we got ready to deal with the elements, some of the snowmobilers stopped by to wish us luck. More of them watched us with a kind of horrified fascination, like reporters watching condemned men. What mostly alarmed them was that we were not dressing for the cold as they did and, in fact, by their standards seemed to be getting ready for a beach outing. However, our needs were much different from theirs. Like ice fishermen and goose hunters, snowmobilers go out in very cold weather, but once they get there they don't do much except sit. Their problem is to conserve every bit of body heat they can. Therefore, they sensibly encase themselves in heavily insulated boots, suits and helmets which, in effect, form flexible thermos bottles. If a snowshoer or any other physically active winter sportsman were to do likewise, he would first be immobilized by the moonwalk-type suit, and if he could move, he would do so with considerable discomfort, even danger.

No matter what the temperature, thrashing around produces a lot of body heat and moisture, and it is imperative that this be controlled and disposed of properly or it will turn to ice with all sorts of evil consequences.

If you are going to be active in the cold, the best mode of dress is a loose, layered one. Fishnet underwear, which provides good insulation because of the air spaces but lets heat and moisture radiate away from the body, is a good foundation garment. Wool makes the best next layer because, unlike cotton, it dries from the inside out and breathes (passes moisture) very well. The outer layer should be thin, designed to keep moisture and wind out. Overall, a long flowing *cagoule*-like mountaineering parka is good. Under these layers, heat and moisture rise, and the top of the costume, around the hood of the parka, can be easily opened or closed to accommodate this. When you start building up too many BTUs and too much moist air, you can open the throat and let it escape, and when you have cooled down, you close the throat and start building up the heat. For the feet, rubber boots with leather uppers are the best.

This is how we remained warm, more or less, through the worst week of winter weather anyone around the Black Forest could remember. As to questions about where was all our stylish, puffy, down gear of the sort smiling models display on the covers of outfitters' catalogs, it was where it belonged, stuffed in our packs, waiting to be used at night when we stopped and when we slept. There is not much that provides better and lighter insulation than down because of its properties for trapping air between the little feathers. However, despite all the down being sold in such arctic areas as San Diego and Houston, it has severe limitations. If it is squashed flat, say, under a pack, it can't trap air and does not insulate. For the same reason it loses its clout if it becomes wet. Down is great for sleeping in or standing around in, but otherwise it is not so hot. In a recent spree of false role playing, and as a status symbol, it has been misused, oversold and overbought.

Snowshoes

On the right kind of trail you can lope along on snowshoes, getting a pleasant, bouncy, trampoline effect from the webbing. You can bound down an open slope in big leaps, achieving somewhat the sensation of rappelling without ropes. Otherwise, there is nothing very sporty about snowshoeing, which is really nothing more than awkward walking. Also, it is not a high-skill activity. Anyone who can simultaneously walk and chew icicles out of his mustache can learn to snowshoe passably in half an hour or so. Like other ordinary things such as crawling, planting bulbs or sawing wood, it gets easier the longer you do it, but the basic act is not difficult.

Snowshoes have been around for a few thousand years and nobody has yet invented a ski, sled or snowmobile that better accomplishes what they are intended to accomplish—getting a man through deep snow and over rough terrain. Because there are a lot of different snow conditions, a great variety of snowshoes have evolved. There are, for example, almost circular ones, often called bearpaws, which while a little awkward are great supports in very deep, soft, dry snow. At the other extreme are long, narrow six-foot shoes with extended wooden tails on which you can travel very fast in hard snow and open country, say, across prairies or tundra. Both pairs of my shoes are of a general type developed and used in woodlands. They are less round and provide less support than bearpaws, are rounder and not so fast as the narrow, arctic type, and have short tails and are more maneuverable than either of the extremes. The larger pair is laced with rawhide and the other with neoprene ribbons. Much as I hate to admit it, the neoprene is more practical because it does not take on water and does not have to be periodically varnished, but I have always preferred rawhide and, somehow, it feels better underfoot.

Sentiment aside, Sam and I thought that we might have snowshoe problems in our travels in the Black Forest. The larger rawhide shoes will keep about 240 pounds on the surface of the snow. The neoprenes are rated at about 200 pounds. Obviously Sam had to use the bigger ones, but with a pack he was going to be 20 pounds or so overweight, and I was going to be about as much over the best load limit on the neoprene shoes. If we got into very deep, light snow we might have some hard going—as in fact we did.

There are a series of open meadows extending from Potato City back into the forest. The wind had graded and packed these snowfields, and we moved across them smartly, cruising along on the crusty surface. However, there was only about half an hour of this easy going. In the woods there was a three-foot snow cover on the flat, but frequent and extensive drifts were twice as deep or more. Also, because it had been so cold, the snow remained very light and fluffy. On our overloaded shoes we sank six inches to a foot with every step. Obviously this was better than sinking in three to six feet, as we would have if unshod, but it still promoted slowness and exhaustion.

Woodland shoes have an upward tilt on the prow. The purpose is to keep the shoe pointing up through the snow, preventing the tip from digging down diagonally into drifts and stopping or tripping its wearer. In effect, these shoes are flattish scoops. Even properly weighted you expect to sink a few inches, and with each step some snow collects on the scoops. Getting rid of it accounts for the customary snowshoe gait, which is more a glide than a stride. On each step the foot is shuffled forward and then given a little shake to kick off the snow. When you have a six-inch dollop of snow to clear, as we did most of the time, you have to lift your foot higher to get out of the hole you have dug and then kick harder.

All of these problems only really concern the first person in a party because he packs down the snow, and anyone who follows can step in his tracks and ride along nicely on the surface. In the flat, better-packed snow, one of us might be able to break trail for a quarter of a mile or so. In the drifts, ascending and descending the mountainsides, we sometimes could keep going for only 20 yards before we were gasping and heaving and had to stop to open our parkas to let off steam and make way for the other to have a go at trailbreaking. It is a situation in which strength is far more valuable than expertise, and as the days went on Sam took longer and longer turns in front. It was the horse's edge—or penalty.

Deer

We labored in the snow, but no more so than the only other mammal much in evidence, the white-tailed deer. There were a lot of deer trails and they were cast in an odd pattern—four small holes, where their narrow hooves had disappeared, were divided by a shallow trough, the mark left as they dragged their bellies through the snow. Generally, the deer were wandering from one big tree to another. On the lee side, where the snow was shallower, they would sniff for edibles—dried grass, shoots, acorns—and if they found them they would paw away the snow.

Frequently a deer would pop up no more than 10 or 15 feet away from behind a log or tree where it had been attempting to feed or escape the wind. It would make a few lunges through the drifts in a kind of token escape effort and then stand heaving and exhausted, staring at us dispiritedly, much more concerned and bothered by the winter than by two silent snowshoers.

A rap against snowmobilers has been that some of them, finding deer in these situations, have delighted in chasing and harassing them. Besides being plain ornery, this can be destructive of deer. They are incessantly searching for sufficient food to fuel their body-heating plants. Anything—like escaping a carousing snowmobiler—that requires them to burn up a lot of extra fuel, diverting it from the main job, decreases their chances of making it through the winter. Fortunately, this sort of cruel and stupid hazing is not as common as it was when motorized sleds were more of a novelty. Now when such jackassery is encountered, it is treated harshly, not only by game managers but also by other snowmobilers.

When we met deer at close quarters we passed by very quietly and carefully, trying to reassure them that we had no hostile or antic intentions. Like them, we were essentially concerned with survival.

One of the tenets of pop natural history, conservation and ecology is that all creatures have evolved so as to be perfectly fitted, if man will let them alone, for the niches in which they are found. Tidy as this theory is, it is often not true, as the problems deer have in winter demonstrate. Whitetails are, in fact, badly equipped for a Black Forest winter, a niche which they are often in and with which they must cope. They cannot travel through the snow like hares; they cannot rise above it like squirrels; they cannot store food like beavers or tunnel after it like shrews; they cannot sleep through the cold like bears or bats.

"If evolution worked," said Sam, staring at a deer obviously depressed by the cold and virtually immobilized by the drifts, "they would grow and shed snowshoes instead of antlers."

Only in the very long range does the situation of the deer make sense. The fact that deer, like a lot of the rest of us, are not well equipped to deal with fairly ordinary environmental conditions is the bottom-line control on the size, location and activity of populations. Worry about this is very fashionable these days, but as yet no environmentally dominant beast has been invented. Each in our own way, we all remain flawed.

Cold Camps

Our plan was to follow, for as long as we could bear it, the Susquehannock Hiking Trail, a blazed footpath that makes an 85-mile loop through the Black Forest. The blazes were prominent enough, being mostly above the drift line, but the trail itself was indistinguishable from any other part of the forest. Nevertheless, for a time we stuck with the submerged path because if it offered no better route than any other, it was no worse. In the afternoon we came down the Susquehannock Trail off a long, flat ridge and began to descend a narrow ravine. There the blaze line slabbed along the side of the slope and the footing was miserable. The snow was so powdery that our downhill-side shoes were constantly giving way and we would either toboggan or fall. If you are alone and go down into a drift wearing snowshoes and a pack, getting up again can be difficult. If you have really sprawled, it usually means baring your hands, fumbling around in the snow and unfastening the shoe harnesses, then wallowing upright and, while standing waist deep in snow, putting the shoes back on your boots. Invariably it leaves you colder, wetter and much madder than you were before you fell. If you are traveling with a partner and he is still standing, you can usually claw your way upright, using him for a handhold.

After a few falls, it occurred to us belatedly that there was no law or logic which required us to continue slipping along the side of the mountain following a nonexistent trail. So we slid down the ravine to the bottom where, presumably, a small stream still ran under the snow and ice. There were some drift rows, windfalls, and occasionally we would break through ice arches and get into the watery slush below them. However, in general the stream course was level and an improvement over the mountainside. An hour or so before dark we came to a wide place in the ravine that recommended itself as a campsite because there was a pool of open water—the first we had seen that day—where a strong spring joined the stream. Curiously, a pair of middling-large trout were disporting themselves in this bathtub-sized pool.

You can stay warm while camping in extreme cold, but it takes constant effort and planning. The cold can never be disregarded. It affects and complicates everything you do. As soon as you stop traveling, the rate at which you produce heat declines, and any moisture on your person will begin to freeze unless you put on the down-insulated gear. This slows the freezing process but also makes you more cumbersome just when you have more intricate work to do. Hands are a big problem. There are a lot of things that cannot be done while wearing mittens or gloves—laying tinder, rigging a tent, untying frozen knots on food bags, some cooking and eating. Yet when it is colder than 10° below, as it was, it is painful, even dangerous, to work bare-handed for more than a few minutes at a time. You pick at a job until your fingers turn stiff and awkward. Then you put them back in mitts until they thaw a bit before you get on with the job. We use a three-man backpacker's tent, which ordinarily we can put up in less than five minutes, but in the cold it takes 15—and they are 15 hard minutes.

Beyond making ordinary jobs slower and more difficult, excessive snow and cold create a lot of new ones. You cannot simply find a level spot and pitch a tent. First you have to hollow a cave out of a drift (as a windscreen and for insulation) and then tromp around on snowshoes to make a floor. Before you build a fire you have to scoop away as much snow as possible and then, on top of what is left, lay a raft of logs on which your fire will float, so to speak, in the snow. If you are careless about putting down a knife, a plate, a bootlace, it is very likely to disappear, either because it sinks into a drift or is covered over with new snow. (It snowed steadily the first day and intermittently the next three. The snow that fell was continuously supplemented by snow blowing off trees and from the ground.) To be fair, there are compensations, the major ones being that under these conditions you are seldom troubled by mosquitoes or nettles.

There is also a real sense of urgency about living in the open in very cold weather. Even with a good fire (and winter ones are often not the best), you cannot lounge about a camp. It is not good to sit down because of the moisture-to-ice factor. Everything—your breath, nose drippings, bread, a plate of boiled rice, a cup of coffee—freezes very quickly when it is removed from the immediate vicinity of the fire. You cannot take much time with any job—mending a snowshoe harness, for example—because if you do, you and the material are apt to freeze.

"I've been colder," Sam said as he was punching new holes in a very hard harness strap, "but the thing about this is all the work it takes to keep from getting cold. You can't relax. It's always waiting to come in."

Sleeping Cold

About the only place where you can escape the cold for any appreciable time is inside a tent and a good sleeping bag. Much of winter camping is a race, the object of which is to get into this position as quickly as possible. Like so many other things, however, it is something of an obstacle course. The first law of going to bed in the cold is a negative but important one: take no snow into the tent. If you do, it will first melt, then soak your bag and finally turn to ice. In consequence, the 10 minutes preceding retirement are devoted to scraping ice and snow off everything. Then you ease inside the tent as gracefully as a stiff, tired, padded person can and sit on the tent floor with your feet hanging outside. So reclining, you pick away at frozen knots and in God's good time may be able to remove your boots. These are brushed free of ice and snow and stuffed into the foot of the sleeping bag. Heavy boots may be a bit knobby, but some nighttime knobbiness is preferable to spending 15 minutes the next morning beating on your boots with a stick to loosen them up enough to put them back on your feet. Anything else you are not wearing, but would like to wear in the morning, is put under the bag. Whatever it is, it will not be warm in the morning but it may be flexible. Finally, you put on a hat and dry socks, and squirm into the bag. You pull the drawstring tight around your muzzle, leaving only enough of an opening through which to breathe.

All of this is done in the dark while hunched over in a small nylon shell and while one is under considerable pressure to hurry. If you have removed one boot, you are well motivated to get the other one off quickly and both feet into down insulation. There is also some social pressure for speed. If you are the first one in the tent (there is no way that two people can manage this getting-ready drill simultaneously), you are spurred on by sharp queries—"What the hell are you doing now?"—from whoever is shivering outside by a dying fire. If you are the second one in, you are harassed by muffled orders which come from inside the cocoon—"Tie down the flap." "Hey, you got your knee in my stomach."

Sioux Alarm Clocks

Outside of such improbable occurrences as the coming of an archangel, there is only one thing that can make a person who has worked his way into a sleeping bag get up in the middle of the night, go outside into $-20°$ temperature, remain there briefly and then repeat the whole tedious getting-in-the-bag process—all of which brings up the problems of water in very cold weather. They are various and curious.

Our first camp was the only one where we had normal, that is, liquid, water. Thereafter, to get a drink, or for cooking at night, we had to melt snow over the fire. Again, this complicates and extends what should be a very simple act. Also, snow water tends to taste like old leaves, and nowadays there is always the unnerving suspicion that it may be much hotter than it seems by reason of a lot of invisible particles that have wafted in from northern China.

During the day we slaked our thirst, which was as real as if we had been packing in July, by gobbling handfuls of snow. But this never seems to put enough moisture back in your boiler. At the end of a cold day you are always a bit dehydrated, and one of the biggest rewards of making camp is getting two or three drinks of real water, even if it means boiling snow. However, whatever goes in must come out.

I once had a friend, a professor of physiology, with whom I did outdoors things. He had given a lot of serious thought to water control and was the first to tell me about and demonstrate what he called the "Sioux alarm clock." He claimed that the Sioux, by adjusting their water intake at night, could leave themselves a morning call for any desired time. He had experimented with the technique and by the time I knew him could, indeed, awaken himself within 15 minutes of a predicted time.

This is important for any very cold camper. You want and need a lot of water when you get into camp, but thereafter it is prudent to go easy. At the tag end of the brief evening, even though nothing else is standing between you and bagging it, it is well to hang around a while, endure a little more freezing for the sake of stabilizing your water system. Nobody minds getting up too late in the cold, but the prospect of the old Sioux alarm clock going off too early is terrible.

Ultimate Drifts

The place where we first camped was only about a mile from the mouth of the ravine where it intersected a larger valley and a small river. However, a few hundred yards below our camp the ravine veered and became exposed to what had been the prevailing westerly wind. In consequence, the narrow gorge had become a single long drift, seldom less than five feet deep, often more. It took us more than two hours to wallow this mile. When we got out we were caked with frozen snow, half-blinded and fully exhausted. There was a strong temptation to simply lean against a tree and cry, but as usual the moisture-cold factor made this self-indulgence undesirable.

Without putting any gloss on it, we were convinced that we had more escape and therapy in drifts than we really wanted and, in truth, about all that we could endure. Therefore, we turned to our topographic map and began thinking seriously about alternatives. What immediately caught our attention was a web of snowmobile trails crossing the forest. We reasoned that even if our friends from Potato City had not been getting out much in the blizzard, these trails had probably been traversed oftener and more recently than the standing drifts and should therefore provide somewhat easier passage. We headed across the valley to pick up the nearest of these trails, and when we did our assumption proved to be true. For the next three days and 25 miles we slogged along on these sled runs. They were blown over in places but always provided better footing than the abominable ravine. Following them was easier and also gave us a chance and time to see something of the winter other than the inside of a drift.

Snow Swimmers

Beside ourselves and the deer and an occasional raven that flapped by, croaking hoarsely and presumably looking for dead deer, the only creatures that seemed very active were porcupines. Now and then we would see trees where they had been feeding, gnawing away bark. We wondered and talked a little about how such a dumpy, short-legged, groundhog-like animal was able to get around in the snow. One afternoon when a porcupine ambled across the trail to a big evergreen 30 yards away from us, we had a chance to find out. Porcupines, it turns out, get along very nicely and easily in the deepest snow. They do so by floating on it, more or less. The big quill-covered tail and the mop of coarse body hair is all extended, giving them great buoyancy. So supported, they dogpaddle with their short legs and move across the surface without sinking more than a few inches. Astonishingly, the fat, slothful porcupine handles deep snow much better than deer, which are noted for their grace and agility.

"He is all snowshoe," remarked Sam of the porcupine we watched. But the porcupine, like everything else, is far from perfect. Whether in the snow or on hard ground he is very slow of foot. Even burdened as we were, we could have overtaken this one at any time and worked whatever our will was on him. Despite its quills, which are not as formidable in fact as they are in myth, the porcupine is especially defenseless against any sort of aggression or predation. On account of this vulnerability there used to be a rule of the woods (now seldom invoked because times and human activity have changed so much) that these animals should never be needlessly molested because they are among the few creatures so helpless that a man in trouble, lost, without tools or weapons, can easily catch one.

Winterism

On the last day we were feeling chipper about things: about having beaten through the drifts, about watching a porcupine swim in the snow, about having made out as well as we had. This is the way with most endurance-type ventures. The doing of them is more agonizing than it is exciting and does not provide many highs, like riding fast on a horse or a snowmobile, running a good rapids or hitting a ball hard. The pleasure is retrospective and comes when you can be pleased about having done some wild thing that you were not sure you could or should do. But this is not all. Almost always there is a curious, very difficult to explain, sensual pleasure, which is a unique by-product of all the hanging on and heaving. The last night out there was a feeling that may suggest something about this phenomenon, if not define it.

We had been overtaken by darkness and fatigue and had made a bad camp on a ridge top. The wind there was especially strong and the snow devils ferocious. The trees, mostly big black cherries, did not offer the protection that hemlocks in a hollow will. The fire was balky and smoky. All the camping chores were aggravating. However, there was a splendid drift into which the tent could be wedged, and once we got into it the night was no worse than usual, just much noisier.

To say that a wind howls may be a cliché, but on that ridge the wind did howl—like God's own arctic wolf, like a hysteric, like an accident victim. Before the howling wind the big trees gyrated, groaned and creaked. Occasionally there was a sharp, riflelike report as a branch or trunk broke. Now and then there was another peculiar sound like a mighty mousetrap snapping shut. The only explanation that we could think of was that we were hearing living wood crack and split as it was squeezed ever tighter in the vise of frost. The moon was up and bright, though not much warmer-looking than a disc of ice. The moonlight filtered through the tent and was diffused by the layer of frost crystals on our fragile ceiling. Through the nylon, backlit by the very cold moon, the weird wind-whipped dance of the trees could be seen. Once in a while the moon would be shadowed and obscured as a snow squall blew past.

It occurred to me that perhaps no one else had ever seen or heard exactly the same things, that quite possibly no one else had ever spent this kind of night on that ridge, that assuredly nobody else was lying out in the Black Forest that night. But beyond this and the spectacular audiovisual display was something else: a powerful, satisfying, in-the-present sensual reaction. The best I can do in words is to say that there was a sensation of being inside winter, not trapped there but simply being in a very exotic and foreign but stimulating place. I do not claim that I would like to stay there long, but it seemed a very fine place to visit. The recreation—the escape and therapy offered—compensated for the unavoidable inconvenience of getting there.

The Original Dude

Though there were never more than 500 of them and they flourished for only about 25 years, the mountain men—the Rocky Mountain fur trappers of the 19th century—have had a mighty impact on the American imagination. They didn't appear until the mid-1820s, but by 1840 dozens of books and innumerable magazine and newspaper articles had publicized (and often exaggerated) their adventures in the theretofore unknown regions of the Wild West. They were known far and wide for their physical prowess, feats of derring-do and go-to-hell lifestyle. By 1838 there was even a kind of mountain-man hall of fame, a waxworks museum in St. Louis in which mountain manikins permanently wrestled stuffed grizzly bears, dueled with gigantic savages and pranced in their exotic finery.

Today, more than a century later, names like Jim Bridger, Kit Carson, Joe Walker and Tom (Broken Hand) Fitzpatrick evoke an era of American history in which freedom and the pioneering spirit seem to have blossomed most fully. In fact, thousands of would-be mountain men still get together at various spots in the West each summer to dress up in beads and buckskin, show off old muzzle-loaders and live in tepees. What these latter-day types are re-creating is the traditional rendezvous, which served the old-time trappers as both an annual business meeting and a blowout.

The rendezvous was the social event that cemented the legend of the rip-roaring mountain men. Early each summer, merchant hustlers from Missouri would come out with supply caravans to meet the trappers at a prearranged site on the eastern slope of the Rockies. There the suppliers would buy furs at rock-bottom prices—a beaver pelt that brought $4 at the rendezvous would be worth more than twice that in St. Louis—and sold staples, coffee, salt, guns, shot, powder, tools, traps and other necessities, at prices marked up as much as 10 times. Essential to this commerce were the bulky kegs of very raw whiskey distilled a few weeks before the caravans started west. The result of this trade was that the merchants got rich and the trappers remained in the mountains, usually in hock for next year's catch. The other crucial ingredient to the rendezvous mix was the Indians. Those with whom the traders and trappers were not at war were invited to the festivities to sell their furs, to participate in the fun and games and, not least, to share their women.

For a month or so, which was about as long as even the hardiest could stand the pace, about a thousand red and white residents roistered about in a mountain meadow some 1,200 miles beyond the reach of the law, property owners, innocent bystanders or any other pillars of respectability and authority. The men whooped and hollered, fought and sported, chased and caught women. When the bacchanal was over, they went back to the hills and the beaver, broke but with their spirits refreshed. In the process, they set impossibly high standards for all subsequent stag parties, fraternal conventions and college reunions.

The first rendezvous was held in 1825 and the last, a dispirited affair, in 1840, when the bottom had dropped out of the fur market. According to connoisseurs of such goings-on—and there were some formidable men who attended and survived nearly all of the rendezvous—1833 was the vintage year. Participants had especial need to frolic that summer because the previous year had been the hardest in the annals of the mountain fur trade. All winter ferociously competitive companies of trappers had been breaking up each other's camps, hijacking furs and generally harassing each other in violent and ornery ways. There had been a lot of trouble with the Blackfoot Indians, as a result of which about 100 white and red men had died. Finally, the weather had been terrible. Blizzards had commenced in October and the snows did not melt until the end of May. In consequence, by late spring all hands and races were eager to let bygones be bygones and get some badly needed R & R. In June, 300 white trappers and 600 or 700 Indians, mostly Shoshones, Nez Perces, and Flatheads, met at the designated site on the upper Green River, 60 miles southeast of the present site of Jackson Hole, Wyo.

Among those there to meet them was William Drummond Stewart, late of Murthly Castle, Perthshire, Scotland. It was his first rendezvous, and he was to remember it as his best: "It was the last good year, for with 1834 came the spoilers—the idlers, the missionaries, the hard seekers after money." Stewart himself was an outsider who was 38 in 1833 and came to the rendezvous for his own amusement. However, he was a tourist of a very special kind: so well equipped by temperament and experience for blood sports and violence, so handy with horses, knives and guns that he made even the trappers sit up and take notice. Later, after Stewart had cut some fairly spectacular notches for himself, one of the trappers was to muse over this strange foreigner: "an Englishman. . . . Well, them English are darned fools; they can't fix a rifle any ways; but that one did shoot 'some'; leastwise *he* made it throw plumb center. He made the buffler 'come,' *he* did, and fout well at Pawnee Fork too . . . what he wanted out thar in the mountains, I never jest rightly know'd. He was no trader, nor a trapper, and flung about his dollars right smart. Thar was old grit in him, too, and a hair of the black b'ar at that."

William Drummond Stewart was the very first of a type that would become an aberrant part of the history of the American West—the hair-on-the-chest, deep-breathing, supremely adventurous (yet always slightly-slumming-it) gentleman sportsman: John Wesley Powell, Mark Twain, Teddy Roosevelt, Ernest Hemingway. All went west in search of hunting trophies and manly excitement—moments of truth and such. Few of the later thrill-seekers found more of them than this doughty Scot, and perhaps none had less tender feet.

Evelyn Waugh could have been describing Stewart when he wrote, " . . . younger sons were indelicate things . . . it was their plain duty to remain hidden until some disaster perchance promoted them to their brothers' places." Stewart, the second son of Sir George Stewart, 17th Lord of Grandtully, was born in 1795 at Murthly Castle. Among many other properties, the 32,000-acre estate included Birnham Wood, which some years earlier had so astonished Macbeth. William and his older (by 14 months) brother, John, to whom the family lands and income were entailed, apparently despised each other from their nursery days. In 1812, having no immediate use for the high-tempered William, Sir George purchased a military commission for his 17-year-old son. William was soon posted to Spain, where he served in Wellington's Peninsular campaigns. He fought very well at Waterloo, was decorated for gallantry and promoted to captain. Thereafter he and many of his well-born colleagues of the officers' corps were demobilized as half-pay pensioners. Having no skills but martial ones and an aristocratic disdain for acquiring civilian ones, many of these younger brothers began roaming the world, waiting for the lucky disaster that might bring them titles and fortunes.

In the 1820s Stewart was a mercenary in Portugal and Italy, hunted in Turkey and the Russian Caucasus and then holed up for some time amid the pleasures of the mysterious caravan city of Tashkent. Following the death of his father, he returned to Scotland and immediately stirred up family trouble. He happened on a beautiful serving girl, Christina Stewart (of no known or admitted relation), who, with her skirts tucked up above her knees, was doing the wash. Stewart, it has been reported, immediately "fell in love with her nether limbs."

Shortly, the leggy Christina became pregnant, which wasn't particularly shocking until, after the birth of a son, William married the girl. This impropriety created a great scandal. The marriage lasted for 25 years, ending only when Christina died, although Stewart kept his wife and son, who was named George, in a separate establishment. He and Christina seldom met, and though their relationship was friendly and he never disavowed her, the marriage hardly seems to have been a conjugal one.

The matter of Christina added further fuel to a flaming sibling quarrel about money. Bickering between William and John, the new baronet, was constant. Eventually the two had such a violent confrontation that it was said the ancient walls of Murthly trembled. William stormed out, shouting that he would never again sleep a night under that roof. And he didn't, but it was a vow that was later to cause him some considerable inconvenience. Stewart thereupon exiled himself to the uncharted wilds of western North America, territories which were then regarded by Europeans as among the most savage and mysterious on the globe.

Stewart later gave the impression that the miserliness of his brother forced him to travel virtually as a pauper. In fact, he set off in high style. He brought a number of wardrobe trunks containing fashionable gentlemen's sporting outfits—for which he owed several tailors. Stewart's most practical and prized possessions were a matched pair of rifles made by the brothers Manton, the leading London gunsmiths. Mantons—big, beautiful .70-caliber guns—were esteemed as the finest hunting pieces in the world and today have rarely been surpassed.

Stewart arrived in New York in May of 1832. After shipping most of his baggage by water to St. Louis, he procured a good horse and rode west—at a leisurely pace, en route enjoying squirrel and bird shooting in the forests of the Midwest. Arriving in St. Louis that fall, he took a suite at the Mansion House, thought to be the finest hostelry west of the Atlantic seaboard. For the next six months he charmed—or perhaps more accurately, overawed—the St. Louis upper crust with his aristocratic manners and arrogance. "His general conversation and appearance was that of a man of strong prejudices and equally strong appetites," remarked an American acquaintance, who, being a staunch republican, disapproved of Stewart on principle but was, like most St. Louisans, impressed by his style.

Captain W. D. Stewart, British Army, as his calling card read, attended race meetings, shooting matches, cockfights and gambling parties with the gentlemen of St. Louis. He attended their ladies at teas, dinner and gumbo balls, which, despite the name, were the toniest social events in town. But these diversions were scarcely enough to quell Stewart's restlessness while he waited out the winter. He wanted to get at the real fun in the Wild West. At first, he may have planned only a bit of shooting on the plains, but in St. Louis he heard about the mountain men and their rendezvous and determined to make their acquaintance. Luckily, he met one of the few men in the settlements who could provide an alien such as himself with an entrée into the society of white savages. This was Bill Sublette, who was to become Stewart's lifelong friend. Sublette was among the first white men to work in the Rockies, going in 1823 as a beaver trapper. By 1833, Sublette and his partner, Robert Campbell, had prospered enough to win the contract to supply the rendezvous of that year, and they agreed to take Stewart with them.

Stewart left St. Louis in May, riding with a pack train commanded by Campbell. (Sublette took the heavier freight on flatboats up the Missouri and then went overland by another route to the rendezvous site.) The Scotsman had managed to replenish his finances by then—a knack that was to serve him well several times during his American career—by agreeing to chaperone the son of future President William Henry Harrison. Young Harrison, a doctor with a drinking problem, was put into Stewart's care to dry out on the trail, for which service Stewart was paid $1,000.

On such expeditions it was the custom for even the leaders to start letting themselves go to greasy buckskins, flowing beards and matted hair as soon as they cleared the settlements. Stewart, however, had no inclination to go native, and besported himself more or less as if he were a member of a shooting party on a grouse moor. Each morning, after ostentatiously arraying his toiletries outside his tent, he would meticulously shave and perform his ablutions. Among the muleskinners Stewart found a man who had done some barbering and arranged for the chap to trim his hair every few days. Also, he either brought with him or found in the caravan a young New Yorker, George Holmes, of some breeding and probably of good if effeminate looks, because he came to be called Beauty. Holmes occasionally served Stewart as a valet.

All of which understandably gave rise to snickering in the ranks— until the caravan reached the big-game country, where vast herds of buffalo, antelope and elk roamed. There it became evident that Stewart had the best horse in the party; that his Mantons were superior to anything the frontiersmen had ever seen, including their treasured Hawken rifles; and that he had the skills to match his equipment. Stewart was later to write that hunting on the plains was a sportsman's dream, but at the time he could not resist a certain nonchalance. The buffalo, he suggested, were rather impressive by reason of their size and numbers, but a bit unaggressive for his sporting tastes. To compensate, he began taking them Indian-style, running his horse flat-out for five miles or so across the open prairie until he drew abreast of one and brought it down with a shot in the ear. He also made repeated attempts to run down antelope, which he admitted moved like "streaks of lightning." In this he was unsuccessful, but said that if he had a decent English hunter (a horse trained for jumping and pursuit), he would wager £1,000 he could turn the trick.

In the Laramie Mountains of what would later be Wyoming, a party of French-Indians who served as the caravan's meat hunters surprised a grizzly sow with a cub. The bear proved aggressive enough to satisfy even Stewart. The hunters fired on and hit the animal, but their light muskets had little effect beyond causing the enraged grizzly to charge and scatter them. Hearing the commotion, Stewart galloped up, sat his horse down on its haunches and dismounted. The bear wheeled and came toward him like a big, bloody, runaway locomotive. Waiting until the roaring animal reared up over him, Stewart dispatched it with a single shot from his heavy Manton. It was an act of skill and daring that more or less instantly earned him his frontier spurs. (Later Stewart and his 6-foot grizzly were re-created in wax for permanent display in the mountain man museum in St. Louis.)

After the Laramie bear episode, Stewart probably could have worn lace on his drawers without being mocked. In fact, he committed almost as outrageous a sartorial act. On the morning the caravan arrived at the rendezvous site on the Green River, he retired to his tent with a bundle that hadn't been opened during the crossing of the Plains. Stewart emerged resplendent in a white leather hunting coat, soft ruffled shirt and a pair of trews, the close-fitting trousers of the Scottish gentry, cut by a London tailor from the green, red, yellow and royal-blue Stewart hunting plaid. On his head he wore a spotless, broad-brimmed Panama. It was reported that even Robert Campbell, no country bumpkin, was dumbstruck by the costume.

Stewart's intuition about rendezvous finery was right on the money. These were indeed occasions when everyone put on the dog, flashed and pranced with their beads, quillwork, bones, feathers, furs and scalps. Among these savage fashion plates Stewart was as much a marvel as they would have been in Piccadilly. More important, he had won their respect. Not only at the rendezvous of 1833, but also in his subsequent dealings with the mountain men, he was to demonstrate an unfailing rapport with them. There is a sense that the wild whites and reds of the mountains were the only Americans whom Stewart accepted as peers. In any event, when he strutted down to the Green River in the plaids of his ancient clan, Stewart commenced what he was to remember as one of the best months of his life.

In trying to imagine these free-form happenings, it's important to keep in mind that nearly all the participants were roaring drunk from beginning to end. In truth, the mountain men may not have had a greater weakness for booze, or drunk more of it, than their contemporaries back East. (As a matter of statistical record, the average whiskey consumption in 1830 was half a pint a day per man in America, a tippling rate about three times our current one.) However, for logistical reasons the trappers generally had to ingest all of theirs during the few weeks of rendezvous. And the quality of what they drank was as formidable as the quantity. The traders always packed in exceedingly young whiskey, which upon reaching the rendezvous was often fortified with gunpowder, pepper and salt. One thoughtful entrepreneur improved his hooch with rattlesnake beads, but when the boys saw them at the bottom of the first barrel, they were less than pleased. The trader was stripped, thrashed and put out of business.

Gambling was nearly as integral to the rendezvous as drinking, and for much the same reason. These were men who lived a lonesome, perilous existence except for the one summer month when they came together. Old sledge, euchre and backgammon games, some operated by Missouri sharks, were played on blankets around the camp, but the big draw was an Indian favorite, the hand game. Somewhat like Chihuahua red dog or Australian two-up, the hand game was simple, but the passion and handle were impressive. A man held a small piece of bone (usually a polished and intricately carved section of the femur of a fox) overhead between cupped hands. After certain gyrations and incantations, he extended his two closed fists toward the players, who wagered against him and each other as to which hand held the bone. Whites played it somewhat like craps, fading, betting among themselves on single turns and sequences, say two lefts and a right. The Indians, whose fondness for gambling was perhaps even greater than their taste for whiskey, often played in teams. They would throw virtually all their communal possessions—from horses to children— into the pool.

The stakes and intensity impressed even Stewart, who, it seems, had at times helped support himself as a gambler. One day, in a howling crowd of bone players, he (or, more precisely, the clearly autobiographical hero in one of the two novels he subsequently wrote) was ogling a Ute girl who was having a bad run of luck. She had put literally everything she owned—domestic wares, jewelry and clothes—into the pot, retaining only a short doeskin undershirt that scarcely covered her upper body. She lost. Stewart, a longtime nether man, reported her "wistfully looking on, her small hands clasped over her beautifully formed limbs, crossed one over the other."

Also observing was Kit Carson, one of the great gallants among the trappers. Carson was on a hot streak and tossed the stripped Ute girl a string of beads. These were more or less legal tender in the mountains and would have been at least sufficient for her to have reclaimed a skirt. However, "with an almost imperceptible look of thanks for the gift, she flung them down again to be risked in the chances of the game."

Field sports, so to speak, included steeplechases, shooting matches, knife and hatchet duels, catch-as-catch-can wrestling and team rumbles. "The acme of accomplishment," wrote a commentator on these forms of wrestling and brawling, "was to throw one's antagonist down and, catching the fingers under the jaw or in the hair, use this fulcrum to gouge the eyeball out onto the cheek with the thumb. If this could not be done, the fighter tried to bite off a nose or ear." It was a rare rendezvous that didn't produce a fatality or two.

Stewart was a conspicuous and well-regarded figure at the festivities of 1833, but he behaved with circumspection, apparently realizing that it wouldn't be good form for a newcomer to be too pushy in this wild society. However, he was involved in one incident that summarized as well as any the spirit of the rendezvous. Along with a typical assortment of thrills—a card sharp was lynched, and there were several eye-gougings—the 1833 party had to contend with rabid wolves. A pack of these diseased animals harassed the camp, adding a kind of Russian-roulette flavor to the proceedings. One evening, having made earlier arrangements with an attractive Indian girl, Stewart kicked Beauty Holmes out of the tent the two of them shared. While trying to sleep in the grass, Holmes was attacked by one of the crazed wolves. His screams created more hilarity than sympathy, and someone suggested it would be amusing to stage a midnight mad-wolf hunt. Preparations for it were squelched by Stewart, who, interrupting his own recreation, emerged from the tent and convinced the trappers that the casualties that would almost certainly result from this drunken sport wouldn't be worth whatever drolleries it might provide.

Strolling around the camp the next morning, Stewart came upon a mountaineer by the name of Joe Meek, who was groggily trying to rise from where the whiskey had felled him the night before. (Meek had come to the mountains in 1829 after beating up an instructor at a school in Virginia at which he was enrolled.) Stewart suggested that, given his condition, Meek was lucky not to have been munched by a wolf. Even with a terrible hangover, Meek had a snappy comeback: "It would of cured him sure—if it hadn't killed him." But whiskey didn't preserve the unfortunate Holmes. He died of rabies a few weeks later and was buried in an unmarked grave along an unnamed stream.

Following his first rendezvous, Stewart stayed in the western wilderness for nearly three years, usually in the company of either Tom Fitzpatrick or Jim Bridger, the two most able partners of the Rocky Mountain Fur Company. The precise route of their travels was never recorded, but it's known that Stewart roamed from the smugglers' town of Taos in what is now New Mexico to the outposts of the British at the mouth of the Columbia River. He saw country and had experiences that previously had been known only to mountain men.

In 1836 Stewart showed up in New Orleans, where he took rooms on Bourbon Street. Again he dazzled local society and improved his finances. He and some British compatriots began exporting cotton to English mills. His journals don't spell out the business details, probably with good reason, but Stewart's commissions were apparently substantial, because the next year, he was able to go back to the mountains in fairly grand style. He left from St. Louis in the spring of 1837 with Fitzpatrick, who was taking the supply caravan that year to the rendezvous. Stewart brought along a considerable private entourage—10 men employed to look after his comfort and two freight wagons. Along with conventional supplies, his wagons carried chests stocked with wines, brandies, hams, sardines, marmalades, dried fruits and other delicacies that had never been seen west of the Mississippi River.

From the standpoint of future generations, Stewart's 1837 trip was his most important because he took with him a young Baltimore-born, Paris-trained artist, Alfred Jacob Miller, to record their experiences. If there was ever a painter who earned his commissions it was Miller, because Stewart was a demanding, often cantankerous employer. Fitzpatrick had made Stewart the second-in-command of the entire supply caravan, and Stewart began shaping up the frontier irregulars as though they were a regiment of Wellington's hussars. He took special pains to make certain that his own artist Johnny would hold up his end. Miller was expected to take a regular turn at night guard and other camp duties and also to sketch from dawn to dusk. On one occasion, having been ragged persistently about the quantity of his output, the artist, who was no toady, turned on his patron and said, "I would be glad to paint more sketches—if I had six pairs of hands."

On another day, Miller was engrossed in drawing Independence Rock, a towering outcrop in central Wyoming that was to become a famous landmark on the Oregon Trail. Suddenly, he was seized from behind and his face forced down into the dirt. Miller later wrote in his journal that he thought he had been taken by Indians. Determined to meet his end bravely, he lay silent, breathing what he expected to be his last. However, his assailant was Stewart, who had crept up and overpowered him to make the point that Miller must always remain alert.

Miller was well aware of the artistic and historical opportunity that the trip offered and so put up with Stewart and the hardships. But he wouldn't pretend to be a frontiersman. During the course of a long, hard rainy spell, Miller griped about the mud and bad working conditions. "Mr. Miller," chided Stewart, "you should not be downcast by inclement weather. On days of rain I am more exhilarated, if possible, than when the day is clear. There is something to contend against." (Half a century later, young Teddy Roosevelt would lie in the freezing rain on the Dakota prairie and shout over to his guide, "By Godfrey, but this is fun!")

Also in Stewart's party was Antoine Clement, a young half-Cree, half-French trapper who was one of the best shots among the mountain men and perhaps the handsomest. Stewart had met him in 1833 and thereafter engaged him as a permanent companion. He went so far as to make the young man a gift of one of the splendid Manton rifles. In New Orleans, Clement had proved to be exceptionally favored by Bourbon Street society, though, as was the case whenever Clement visited civilization, Stewart was often occupied trying either to keep or get his protégé out of scrapes.

Beyond using him as a model, Miller found that Clement had a remarkable talent that facilitated the artist's work. Using the heavy Manton, Clement could hit a bull buffalo on the horn in such a way as to leave the animal temporarily paralyzed but standing on its feet. Miller was then able to approach the stunned and weaving animal and make detailed studies of it.

One morning, Stewart, Clement and Miller rode off on a private excursion. Some 20 miles from camp, according to Miller, "Antoine and our leader commenced quarreling over some order that had been given but not attended to . . . Both were well mounted, armed with Manton rifles, and neither knowing what fear was, it was a question of manhood, not social position. As they rode side by side, and were not at all choice in their language, I expected every moment to see them level their rifles at each other. . . . While things were in this critical situation but every minute growing worse, as Providence would have it, a herd of Buffalo was discovered. . . . The ruling passion overtopped everything else, off went Antoine at a full gallop, under whip and spur, & in a moment our Captain followed suit. The result in a short time was two noble animals biting the dust, each of the late belligerents in great good humor, and the subject of the quarrel entirely forgotten."

In June, Fitzpatrick brought the caravan to the banks of the Green River near the rendezvous site of 1833. There, it was met by the mountain trappers and 1,000 Snake Indians, then the warrior masters of the central Rockies. In a gaudy pageant of welcome the Indians swept down on the white captains, saluting them ceremonially with their 10-foot, feather-bedecked battle lances. Again, Stewart had obviously given thought to how to make a rendezvous splash. He sent to his wagon for a crate that had been hauled across the plains. Opening it, he solemnly displayed and then presented to Bridger—Old Gabe, the jokiest and most loquacious of the mountain men—a suit of medieval armor, consisting of a steel cuirass and the plumed helmet of the Life

Guards, the most ancient and elite of British regiments. Bridger pulled the armor over his buckskins and mounted his horse. Slowly, at a gait he imagined suitable for a fancy foreigner, Old Gabe rode in medieval splendor between the assembled ranks of cheering red and white warriors—surely one of the most improbable scenes ever seen on the frontier. Miller was to record it in two paintings.

From the rendezvous, Stewart, Clement, and Miller returned to New Orleans. Along the way, Stewart had received some marvelous news—his brother John had died. Another letter confirmed that because John had remained childless, William was henceforth the Seventh Baronet of Murthly.

Stewart didn't rush home that fall to claim his inheritance, but dawdled in the U.S., amassing an exotic collection of souvenirs. When he finally sailed the next spring he was accompanied by a bull buffalo and cow, a half-grown grizzly bear, several deer, a pair of cardinals, a bale of seeds, the roots and cuttings of western plants, two Indians of unspecified tribal connections and a number of Miller's paintings. The menagerie was under the supervision of Clement, now attendant to the new Sir William.

On his arrival at Murthly, Stewart declared that his sleeping quarters would henceforth be in a distant outbuilding rather than in the castle, thus keeping the angry promise made to his brother seven years before. Stewart showed little interest in the affairs of the estate. Rather, he turned to arranging his American booty. Sheepfolds were converted to buffalo and bear paddocks. A section of what had once been Birnham Wood was replanted with spruce, pine and birch from the Rockies. Miller's paintings were hung throughout the castle and new furniture, carved to resemble buffalo heads, with real horns and hooves, was ordered for the main hall.

In 1840, perhaps to shake his American obsession, Stewart took Clement on a six-month tour of North Africa and Asia Minor. Exactly where they went and how they entertained themselves was never recorded, but they ended up in Constantinople. However, if the trip was intended to cure Stewart's homesickness for America, it wasn't successful. Back in Scotland, Stewart, despite the protests of his relatives, sold one of Murthly's oldest properties to a neighbor for the equivalent of a million dollars. He used part of the proceeds to pay off old debts. Much of the rest was spent, in 1843, on his final American tour, which was conducted with truly regal extravagance.

With Bill Sublette serving as field commander and recreation director, Stewart gathered around him some 50 young gentlemen and their servants from the better families of St. Louis and New Orleans for what awed western newspapers referred to as a "grand pleasure trip and hunting frolic." The party included sport hunters, naturalists, an embezzler on the lam (who was discovered and discharged) and various gentlemen seeking escape from booze, boredom or social entanglements. On Stewart's tab, they secured the best horses and equipment; they were provided with new, roomy canvas tents and servants. Stewart's personal quarters would have done credit to a khan or a czar. His tent was a 14-foot-square, scarlet creation. The mattress was made of two buffalo robes and was covered with Irish-linen sheets and a blanket of Russian sable. A Persian rug lay on the dirt floor, and otter and leopard skins draped the tables. A huge brass Turkish incense burner sweetened the air. As body servants, Stewart brought a valet and an odd-job boy from Murthly. And Clement was made the master of the hunt.

Despite—or perhaps because of—these preparations, the trip didn't live up to its expectations. At first, many of the young men, especially the naturalists—Sublette labeled them "the bugg ketchers"—and the hunters, had a fine time. They acted out what they were, playboys on a kind of Outward Bound lark. They had no inclination to suffer great hardships in pursuit of anything. They liked their wine and hot meals. They sang campfire songs, repaired to dry tents and sheets and rose at a decent hour in the morning. But Stewart had them cast in quite a different role. He wanted them to be Oriental courtiers to his caliph and also the hard-bitten mountain men he had known 10 years before. It was a conceit that history had already made into an anachronism.

The first western frontier, of which Stewart had a last glimpse during the mid-1830s, was finished by 1843. That year there were no more than 50 white fur trappers left in the mountains, but about 1,000 pioneer families were slowly rolling across the plains toward the Pacific coast in covered wagons. These men, women and children represented the second phase of the American westering movement. The best and brightest of the trappers, such as Bridger, Fitzpatrick, Carson, and Joe Walker (one of the few true explorers among the mountain men and the subject of one of Miller's most famous portraits), recognized the wave of the future and were riding it as Indian fighters, pilots of caravans and guides for military, surveying and prospecting parties.

Sir William Drummond Stewart, the foreign tourist, seemed to be least able to adjust to this change. He spent the summer and the treasure of Murthly trying to turn back the clock to 1833. As the pleasure tour proceeded, he became an increasingly unreasonable and often savage martinet in a vain effort to make his playboys deport themselves as old-fashioned frontiersmen. The young gentlemen were cowed, but not convinced. The more sullen of them began to speak of their demanding host as "His Omnipotence." After a make-believe rendezvous was staged on the Green River, where the last genuine gala of this sort had been held three years before, about half the party deserted, the bad sports making their way back to Missouri in high dudgeon. Stewart and the good sports followed in late September. Then, in New Orleans, for reasons never explained, Stewart had an emotional and terminal quarrel with Clement. This odd couple didn't meet again.

Stewart, now 48, returned to Murthly for good. The course of the last 25 years of his life was pretty much downhill. His relatives and tenants found him to be a contentious man whose one source of pleasure was his American acquisitions. One of the last of these was a 12-year-old boy, Franc, the son of a former New Orleans business associate. Franc came to Murthly just before the Civil War to have his manners polished and stayed on to become, by all accounts, a singularly foppish and snobbish young man. He was widely disliked, but Sir William doted on, and eventually adopted, the young American. This led to an open break between Stewart and George, his son by Christina.

When Stewart died in 1871, apparently of pneumonia, there were persistent rumors that Franc had murdered him. However, the charges were never proved and Franc did not in fact inherit any of the Stewart lands or fortune. (Perhaps as something of a consolation prize, Franc stripped Murthly of some $40,000 worth of furnishings and decorations and returned with them to Texas. There he passed himself off as Lord Stewart until an enraged British House of Lords exposed his deceit.)

Stewart's final bitter years may have been no more or less than he deserved; he was a man whose character determined his fate. Yet, somewhere in the western wing of our gallery of folk history there should be a special niche for him. The Seventh Baronet of Murthly was the original American dude, the flashy, ornate mirror in whose reflection we see our wild and woolly heritage.

Messing-Around Rivers

A very strong beginning is one reason for the enduring popularity of *The Wind in the Willows*. Mole, who is suffering from ennui, wanders off one morning from his diggings and comes to a running river. There he meets Water Rat, who invites him to go for a paddle in a punt. Mole is enchanted and admits that he has never before been on a river or in a boat. Shocked at such cultural poverty, Water Rat remarks, "Believe me, my young friend, there is *nothing*—absolutely nothing—half so much worth doing as simply messing about in boats. Simply messing. . . ."

Even if a rat did say it first, I have always thought this was a profundity of, so to speak, the first water, and have acted accordingly whenever and wherever possible. One consequence is that, along with a good many thousands of other citizens, I have gone frequently to the Ozark National Scenic Riverways, a ribbonlike national park in southern Missouri, and have come to think very well of the place. The Park Service being the serious-minded outfit that it is, the Water Rat principle has never been openly acknowledged there, but this is probably the best public property we have for messing about in a boat on a river.

The Current River (from La Rivière Courante, or Running River, so named by French fur traders, who were the first whites to see it) and its principal tributary, the Jacks Fork, are the *raison d'être* for the Ozark National Scenic Riverways, hereafter the ONSR. They rise 25 miles apart in the central Ozarks, south of Fort Leonard Wood, and flow separately through narrow valleys for 35 miles or so, meeting a few miles from the village of Eminence. Thereafter the Current, absorbing the waters of the Jacks Fork, continues into Arkansas and joins the Black, which empties into the White, a tributary of the Mississippi. The ONSR, which was established in 1964 as the first park of this sort, extends from points upstream on the Current and the Jacks Fork, where there is just enough water to float a canoe, to a southern boundary on the main stem of the Current about 30 miles below Van Buren, a community of some 850, where the park headquarters is located. There are 81,216 acres in the Riverways, which is seldom more than half a mile wide, stretched out for 140 miles along the banks of the two rivers. A few small enclaves of private land remain, and the rivers are occasionally crossed by public roads, but, in general, the waters and adjacent lands are undeveloped and free to be used by whoever chooses to do so.

The Ozarks are of great geological age and have been so worn down that the hills along the valleys are less than 1,000 feet high. However, they press in tightly on the rivers, which in many places have had to cut their way through and around the limestone ridges, creating sheer cliffs and gorges that give the illusion of being much deeper than they are. There is an Ozark saying that describes this situation: "The knobs here ain't high but the hollows are awful deep."

Because they rise from no great elevation, the drop in the rivers is a modest seven or eight feet to the mile. There are no big rapids, explosive haystacks or tortuous chutes of the sort that make classic whitewater rivers like the Youghiogheny or Cheat. According to the scale on which canoeists rate rivers—O is essentially flat water and VI is suicidal to impassable—most of the ONSR is classified as I with a few II stretches. This means the ONSR is safe and enjoyable for almost anyone who can tell one end of a paddle from the other. Since the park was opened it has provided for some 20 million hours of canoe use, during which there have been only four canoe-related fatalities.

They may not be risky or rampaging, but the two rivers aren't dull. For one, they are steplike, moving with fair speed down long gravelly inclines that end in pools 50 feet deep or deeper, called holes. This arrangement is repeated for many continuous miles. The drops provide a degree of excitement, and the holes, many of which are at the base of cliffs, are good places just to sit and look about. The Jacks Fork and the Current, at least in their upper parts, are also very narrow with enough twists, turns and sharp corners to give a sense of what it is like to play and be played with by currents.

Some years ago, with the two like-minded friends, I went to the Grijalva, a big, wild river that runs across the Mexican state of Chiapas, just north of the Guatemalan border. Our ambition was to be the very first to take canoes down El Sumidero, a Grand Canyon-scale gorge that the Grijalva has cut through the Sierra Madre. So far as pioneering went, we were disappointed. A day or two into the canyon we began finding empty film boxes—discarded, as we later found out, by a party from Utah which had preceded us.

We were at the bottom of the gorge for about two weeks, locked between the narrow cliff walls that rose several thousand feet. We ran some of the rapids, but drove eyebolts into the walls and roped around more of them. It was horse work. We were exhausted most of the time, worried about the problems the river presented and worried whether we could cope with them. When we finally got out, we were full of ourselves, proud that we had made the effort. However, we didn't have a sense of knowing or caring much about El Sumidero except as an opponent. It was an adversary experience all the way.

In contrast, the Jacks Fork and Current promote intimate acquaintance. They make so few demands, create so little distraction, that there is a soothing sensation of losing oneself in them, of flowing with rather than contending against. To float them is somewhat like meeting a stranger with whom one hits it off immediately and effortlessly.

"Nothing seems really to matter, that's the charm of it," Water Rat said of his river, which must have been quite like these Ozark streams. "Whether you get away or whether you don't; whether you arrive at your destination or whether you reach somewhere else, or whether you never get anywhere at all, you're always busy, and you never do anything in particular; and when you've done it there's always something else to do. . . ."

The Jacks Fork and the Current are fed by hundreds of springs that seep, drip and gush out of the banks into the water or well up underneath it. The largest of them—in fact, according to hydrologists, one of the 10 largest springs in the world—is located a few miles south of Van Buren and is called, not very imaginatively, Big Spring. It flows out of the rubble of a collapsed cavern a few hundred yards inland from the Current and, when it is running strong, discharges 840 million gallons of water a day. By way of comparison, the city of St. Louis uses, on the average, 250 million gallons a day.

There are other springs of note, too. Sometimes our parks are called national jewels, and there may be no one park property that can be better described in this way than Blue Spring in the ONSR. It compares favorably with any geyser, canyon or ancient tree we have. It is a very large spring, flowing at a rate of 90 million gallons a day, but it is its quality, not quantity, that makes it such a jewel. It rises at the base of a sheer, overhanging dolomite cliff and discharges into a pool that is perhaps 100 feet in diameter and, as measured by divers, at least 256 feet deep. Because of this depth and mineralization, the water is spectacularly blue, somewhat the color of the glass once used in Bromo-Seltzer bottles. Despite the color, the water is remarkably transparent; small white pebbles a hundred feet below the surface are clearly visible.

The surface of the spring rolls gently, like a pot just beginning to boil, and the pool is decorated with a garland of watercress. At the outlet of the hole, Blue Spring becomes the equivalent of a respectable trout stream, which flows in a froth for a quarter of a mile before emptying into the Current. A trail from river to spring follows this outlet. The last time I was there, on an unseasonably warm April day, I followed a teen-age couple along the trail to Blue Spring. They were burdened with six-packs of soda, a transistor radio and assorted combs and brushes. Also, they were so entangled with each other that I thought at first they probably didn't care much about what was at the end of the trail. This turned out to be an unworthy suspicion.

After we became acquainted, the boy and girl said they were juniors at a high school 50 miles away and had cut class for the good reason that it was too nice a day to be inside. We arrived at the pool and, without paying much attention to each other, stood staring into it. The effect is like being on top of a deep blue sky, say over Montana, looking through it at the world. After a bit, more or less simultaneously, we each said, "What a beautiful place," or words to that effect. Then the girl gave her friend an affectionate nudge and said, "He's never been here before. I told him I'd show him something real neat."

The springs have great impact on the ecology and recreational activities of the park. Because of them, the Jacks Fork and the Current are canoeing rivers for all seasons, whereas most streams of comparable size usually shrivel up and become unfloatable in the hot summer months or in drought years. The Ozark springs are affected by rainfall, but their fluctuations are slow because they drain subterranean reservoirs that are impervious to evaporation. Some of the subterranean drainage systems are very extensive. Hydrologists using dyes have found that water entering the ground near the town of Mountain View resurfaces in Big Spring 45 miles to the west.

Because so much of it is filtered through rock, the purity and clarity of the water is as fine as any we now have in open moving streams. It is good to look at, or through, and it supports an abundance of aquatic creatures, whose activities are remarkably visible. Brown trout reproduce in the upper sections, and along with smallmouth and largemouth bass, goggle-eyes and suckers are the principal fish for hooking or gigging. Jack salmon (elsewhere called walleyes) aren't so plentiful. There are also thousands of turtles—painters, sliders, soft-shelleds and snappers—lining the banks, basking on rocks and logs. Sometimes it is possible to drift along accompanied by herds of turtles. During the course of a day's float, one is seldom out of sight of green herons and belted kingfishers and recent signs of beaver, mink and raccoon, all of which indicate wholesome waters.

Originally, the adjacent banks and valleys were covered with pine, but these forests were nearly all cut and floated downstream in a logging boom that occurred at the onset of this century. Now oak, hickory and sycamore are the dominant trees, and the jungly woodlands associated with them extend more or less continuously along the rivers. The woods have been well surveyed by natural historians, who have determined that there are some 1,500 species of plants and 300 species of birds. A good many inanimate attractions have also been found and catalogued. For the same reason that there are lots of springs—an abundance of soft, soluble rock—it is estimated that there are from 100 to 300 caves. The ONSR is also thought to be first among midwestern parks in numbers of archaeological sites, there being few campgrounds along the river or shelter caves in the bluffs that weren't used by Indians.

All across the country, concentrations of Indian signs and relics are invariably found in places that are still popular because they are so attractive, interesting or sporty. Long before the plutocrats of the Gilded Age discovered them and began inventing salad dressings there, the Iroquois were taking their summer vacations on the Thousand Islands of the St. Lawrence River. It may not be of great archaeological importance, but I find it comforting and thought provoking that at least in regard to such things as resorts, scenic waterways and nice campsites, man's tastes seem to be independent of time and culture.

There are no bad seasons on the Jacks Fork and the Current, but it is almost impossible to come to them at a better time than in good weather in mid-April. Then some shadblow is still in bloom and the dogwood and redbud are in full flower. The turtles and morel mushrooms are out, but the gnats and mosquitoes are not. The weather is suitable for sunning and swimming by day and for driftwood fires and down sleeping bags by night. Having driven from Arizona with this in mind, I met some friends from the University of Missouri at the place where Shawnee Creek empties into the Jacks Fork, and from there we began drifting.

Along with everything else, mid-April is a very good time in the ONSR for watching Cooper's hawks. At other times of the year, though they are abundant, not much is seen of these hawks as they pursue other birds swiftly but secretively through the heavy woods. (Blue darter is a common and good name for them.) However, for a few weeks in the spring, when they are selecting territories and preparing to nest, pairs of them engage in marvelous aerial dances. The open space over rivers seems to be a favorite place for such displays, and we came on half a dozen pairs. Sometimes they flew so close together that it seemed their wing tips must surely be touching. At other times they broke apart like square dancers, separately pirouetting and looping but always giving the impression that they were responding to the movements of their partners. Together or singly they would spiral up a thousand feet or so and then dive straight down almost to the surface of the water. One of these dances might occupy half a cubic mile of space, and whatever the birds may have been feeling, they gave the impression of great exuberance and elation, of being, in so many senses, very high.

The bowperson with me on this trip was a longtime friend, Tracy Frish Walmer, then a junior at the University of Missouri. Her only known flaw of character is that she was a student of television at that institution. Otherwise, she was quick and curious, observant and knowledgeable about a great number of things. Until that weekend Tracy didn't know Cooper's hawks, at least by name, but when we met them she was enchanted. Being an impassioned bird-of-prey fan, I began carrying on about the migratory, mating and hunting habits of these accipiters, but Tracy may have been the closer student of the nature of the thing we were seeing. "My God," she said, "think what it must feel like to do that."

There are many people who lead useful, interesting lives without being able to tell a hawk from a handsaw, but I think only a few eccentrics wouldn't find it a pleasure to study a pair of cooper's hawks dancing over a river in the spring.

That the Jacks Fork and the Current are high-quality messing-around rivers has been recognized for a long time. Setting aside any conceits about how and why the original Indian inhabitants enjoyed them, it is documented that local white folks have been using these waters recreationally for most of this century. Before nylon, aluminum and freeze-dried beef stroganoff were available, the custom was to load up a wagon or two with canvas tarps, cots, substantial supplies of food and drink, some hand tools and a few stacks of green, rough-cut lumber and drag all of this to a point well up on one of the streams. The first order of business was to unload the planks and slap together several johnboats, which are flat-bottomed, square-ended scows. These would then be sunk in the water, and the recreationists would entertain themselves while the wood swelled and caulked itself. After the boats were judged to be tight, they were raised and dried, and the the party would drift and pole downstream for days or weeks, absorbing, as available, bass, turkey, deer, coons, possums, greens, mushrooms, pan biscuits, grits, pork products and, now and then, liquid corn.

In the '20s, the gentry from St. Louis, Kansas City and even more remote places began to hear about what a lot of fun it was to float down Ozark rivers in this fashion, and a few of the more enterprising residents got the idea that they could probably make money selling this sort of fun to the gentry. Veteran rivermen in the area still fondly recall how flotillas of commissary and fishing boats were organized and crews of guides, cooks and camp attendants were employed when notables like August Busch and Fibber McGee and Molly came to mess around the rivers. Less well remembered because they were do-it-themselves floaters—but more influential in terms of what was to become of the waterways—were some out-and-out nature lovers. Among them was Aldo Leopold, the 20th-century Thoreau, who came to drift on the rivers and went back to Wisconsin to extol them. One of Leopold's disciples was Leonard Hall, who spent years exploring the Jacks Fork and the Current and eventually wrote a book, *Stars Upstream,* which came to be regarded as a natural-history classic. It was also instrumental in creating interest in establishing the national park.

The commercial floating and guide business declined after World War II, but the johnboat is still the craft preferred by locals. Poles and green wood have long since been replaced by outboard motors and aluminum, and in the last five years or so, jet johnboats have appeared. They are driven by propless 50–200-hp motors that sound like the Concorde; they drift, so to speak, at 40 mph, drawing only three or four inches of water, and can therefore run almost anyplace on the lower parts of the two rivers.

Canoes were seldom seen thereabouts during the first half of the century. "They just looked so peculiar and people were scared to death of them," says Hawk Daniel, a veteran waterman, retired construction worker and now a county official who lives in Van Buren, the only community located directly on the Current near the national park. "The general opinion around here was that if a man stepped into a canoe he had suicide in mind," explains Daniel, "but I'd traveled around on construction jobs and I'd seen them. When I settled down here I got to thinking that if people had been using them so long there must be some good to them. I heard about a fellow over in the Bootheel [as the extreme southeastern corner of Missouri is called] who had one for sale. I went over and it looked good to me, and I bought it and I've been using it with pleasure ever since. I got that canoe in 1954 and so far as I know it was the first one on the river around here."

At about the same time, Gaylord (Buck) Maggard, who lived 50 miles upstream from Van Buren, began to think about canoes in a more commercial way. Maggard was then farming, operating a general store and running a cable barge at a crossing called Akers Ferry. "My wife's people had been there for about a thousand years," says Maggard of the establishment, "but they'd begun to peter out. We'd moved out to Colorado after the war, and they kept asking us to come back. Well, finally we did come back and took over at the ferry. In 1955 I bought me six of those aluminum canoes. They just sat there for a good while, and people sort of snickered at me, but then some of the strangers found out I had them and began renting them and then more and more came around and pretty soon I had 300 or so of those canoes, and things got to be like they are now."

How things are now is that there are some 40 liveries in the vicinity of the Jacks Fork and the Current, which among them have some 3,500 canoes for rent. On a hot summer weekend most of them will be rented and in the water, along with hundreds of private canoes.

Renting canoes and selling gas, groceries, drinks, fish baits and sundries to paddlers is now perhaps the biggest industry except for lumber and agriculture in the two Ozark counties that border the river. So far as actually using them goes, Daniel is probably still in the local minority, but nobody in Shannon or Carter county snickers about the canoe business anymore. And what the Park Service is going to do about canoes in the future is the most controversial political-economic issue in the area.

In the mid-1970s Park Service people (The Government, as they are locally called) decided that there were too many canoes on their Ozark rivers and that the paddlers were becoming unable or unwilling to behave as the Park Service thought best. Because no planning agency had devised a formula for calculating the acceptable number of canoes, this decision was essentially subjective. As such, it hasn't been and probably cannot be explained with great precision. However, it seems to me that it reflects the fact that collectively the Park Service has some very strong notions about what is True, Beautiful and Respectable.

My subjective opinions are as follows: The Park Service is a mainstream outfit that has a corporate sensibility someplace between that of Mark Trail and Captain Kangaroo. Though formal statements and reports may indicate otherwise, the Service, in its heart of hearts, doesn't have a high regard for recreation of the unstructured, sensual elative, yelling and hollering sort, and it doesn't enjoy catering to people who do. Rather, the Service sees its mission as interpreting Nature in a noncontroversial way—identifying, pointing out and preserving Beauty and providing the public with wholesome Outdoor Experiences. Individual parks are therefore the equivalent of Lesson Plans, within the framework of the overall course, which might be called Natural Uplift. Each is devoted to one or more clearly defined topic phenomena—canyon grandeur, geysers, very big trees, fossils, ancient pueblos, Civil War fortifications, etc.

It is my impression that the Park Service also has a sharp image of who its customers are—or should be. Ideally, they are a serious-minded family of four belonging to the white, white-collar, suburban class. They think well of the environment but in a responsible, not radical, way, being more supportive of Scouting and the National Wildlife Federation than the Sierra Club or the National Rifle Association. After putting the dog in a kennel and notifying the police that they're on vacation, they come to a park to see the main attractions, as designated by the Park Service. They proceed to them over designated trails and roads. They avidly collect information, especially facts about how tall, deep, old, rare and peculiar objects of nature can be. They are energetic and informal, but deport themselves as people do at church picnics and adult education classes.

By 1975, paddlers were logging over a million of what the Park Service came to call "floater hours" on the ONSR, and the figure increased by 10% to 15% for the next four years. (It has recently leveled off.) A good many of these paddlers, unmindful of the Park Service image, had found free-form ways of messing around the rivers, which were wildly different from those of the model users. For example, word got around among high school and college students throughout the Midwest that floating the Jacks Fork and the Current was a very decent and cheap way to enjoy a weekend. You could convoy down in a van and a couple of Bugs, rent 15 canoes, load them up with cold cuts, tacos, chips, M&Ms and as many cases as you could afford, and without doing much work float along listening to Casey Kasem, dinging on rocks, having water fights and capsizing accidentally or on purpose. You could yell and yodel, skinny-dip and go wherever you wanted. It was also discovered that you can get really weird softball or Frisbee football games going in shallow rivers; that you can body-float trailing along behind an empty careening canoe; that a good way to study turtles is to slap a paddle on a log full of them and watch them scatter; that you can tie three or four stern lines together, carry them up a cliff, tie them onto a tree and swing out over the water, giving Tarzan yells before dropping into a deep pool; that every now and then somebody breaks a leg playing Tarzan but the turn-on is worth the risk; that you can find caves that make cool places on hot days to mellow out in; that redbud and dogwood flowers look good in the braids of either sex.

These are exactly the kinds of attitudes and activities that make the Park Service very nervous. They aren't wicked, but they're irreverent, like giggling during a sermon, and contrary to established ideas of how people should appreciate parks, be uplifted and informed by them.

To set things to right, ONSR administrators first attempted to regulate the number of canoes that livery operators could rent during the busy summer season. The rental agents were enraged by this quota system, and in 1976 one of them, Irby Williams, took the matter to court. Williams argued that because he kept his canoes on private land and put them in the water at public access points, the Park Service, which didn't own the navigable rivers, had no business telling him how many canoes he could rent. The courts agreed, but since then the Park Service has challenged this decision, using a 1980 U.S. Forest Service case as a basis for their new court appearance. In the 1980 case, the courts ruled that a Federal agency has the authority to regulate commercial operations affecting public use of lands and waters administered by that agency.

The idea of rationing canoe use on the two rivers is very much on the mind of Park Service officials. "We are concerned that the quality of the experience will inevitably suffer unless there is some regulation on use," says Art Sullivan, the superintendent of the ONSR. "We're continuing to investigate management techniques for exercising control."

In 1970 a seven-year study aimed at measuring the use of the rivers and the environmental effects was authorized for the ONSR. This is a standard Park Service response when its agents sense an area may be overused. It is certainly not an unreasonable one; in many parks the presence of too many people—and the facilities they require—has already had a bad effect on what land managers call "the resource."

Squads of natural-history specialists were therefore turned loose in the ONSR. The final report, a 139-page document issued in 1978, strongly suggests that the researchers anticipated that the resource was indeed deteriorating because of overuse, and that they looked hard for evidence supporting that assumption. However, little such evidence was found. According to the researchers, water quality remained excellent; it didn't appear that either the status or habitat of any of the flora or fauna (terrestrial or aquatic) was being adversely affected by park users; springs, caves and archaeological sites weren't degraded or vandalized; no severe problems involving visual pollution, littering, trash and waste disposal or general uglification were identified.

The overall conclusion—that the park was in good environmental shape—was a surprising one for a study of this sort and indirectly confirmed the uniqueness of the area. The Jacks Fork and the Current and the narrow valleys through which they run are in a sense a self-cleaning and self-restoring unit, because high water in the spring and fall sweep away accumulated debris. Along the rivers most of the primitive camping is done on bare gravel bars from which all signs and effects of human use are periodically erased.

The research studies also tended to support a long-standing claim of canoe devotees—that theirs is perhaps the most environmentally benign form of outdoor recreation. A canoe makes no ruts, tracks or holes; crushes, tramples, digs up nothing; requires no highly developed trails, roads, launching or docking facilities; leaves behind no odor, noise, dung, fumes or grease spots. Also, besides making fair rain shelters, tables and benches, canoes are good, big trash containers. It's almost as easy and natural to throw a can into one of them as to toss it into the river or bushes.

Along with natural historians, a good many sociological investigators took part in the ONSR study, fanning out to count and interview recreationists. Among other technical devices, they used automatic cameras hidden in bushes. When the propriety of this was questioned, it was explained that the cameras were focused so that the canoes could be identified but the paddlers of them could not.

However, the surveyors did collect some information that documents an interesting phenomenon, which, though not often publicized, exists in many of our national parks. It is that overcrowding is often more a statistical than a real recreational problem. In 1974, the researchers calculated, about 121,000 paddlers had used the rivers. (Today it's estimated that there are 300,000.) Had the canoeists been evenly distributed in time and space, this would mean that on any given day there would have been a paddler every 100 yards along the 140 miles of the waterway. However, this wasn't the case. As a matter of record, 65,000 of the canoeists used *only* a 30-mile stretch of the Current, where there are many access points and developed campgrounds. In contrast, during the same period only 130 paddlers were recorded on the uppermost section of the Jacks Fork.

Some 66% of the canoe traffic was on the rivers during the three summer months, while April and October, probably the best times to run these or any other of our rivers, each accounted for only 5% of the total use. Finally, half the use of the rivers took place on only two days of the week, those being Saturday and Sunday. This all means that if you like floating along in an aluminum logjam you can do so by going to the upper Current on a July weekend.

Dean Einwalter is the park administrator most involved in a day-to-day basis with questions about how many people use the ONSR. He agrees with the findings of the research study that the physical environment of the park isn't immediately threatened and also that the situation has not changed appreciably since the report was issued in 1978. Nevertheless, Einwalter remains concerned about overuse and management techniques for dealing with it. "It's more a psychological than physical thing," he says. "The quality of the experience is threatened because at times and places there are too many canoes in the water. Also the jet boats. There are not so many of them, but they're getting too big. They're inappropriate."

The Park Service is now drawing up regulations to restrict the horsepower of motors used on the riverways. The new jets are thought to be too noisy and potentially dangerous. Jet boaters, of course, think otherwise. "The Government has it in for us," says one of them. "A hippie canoeist from St. Louis can do whatever he or she damn well pleases, but if you're a local out with your family and want to open up a little on a Sunday afternoon, they're all over you. They act like we haven't got brains enough to know what a nice place this is, and we've only been on this river all our lives."

Einwalter says it is difficult to define precisely what constitutes an experience of quality. "I think the park exists so people can appreciate the natural resource in a quiet, contemplative way," he says. "I definitely do not think the mission of the park is to manage a party river or a racecourse for jets, which, without regulations, it's becoming."

It's easy and tempting to be snippy and sarcastic, but in these matters there is plenty to be said for caution and even a certain amount of stuffiness. National parks aren't disposable items like paper plates or plastic cutlery to be used up and thrown away. They were established and are maintained because they are irreplaceable public possessions. There is incontestable evidence that they can be misused and overused, that the natural wonders and beauties within them can be degraded or destroyed by too many visitors. For example, Yosemite Valley. Just because this hasn't happened to the Jacks Fork and the Current doesn't mean it cannot. Therefore, it's prudent to have people like Sullivan and Einwalter watching over and worrying about them.

On the other hand, rating pleasures according to their quality seems to me to be difficult, in all likelihood unnecessary and a bit officious. The same thing—the increasing number of users that alarms park officials—might also be taken as evidence that the quality of experience is viewed as being pretty good. Offhand, and without the benefit of a hidden camera, I can't recall meeting anybody on these rivers who seemed to be depressed because of what they were not experiencing. Most of them appeared to be enjoying the hell out of what they were doing.

It's true that there is something subliminally annoying about the stupid things that interest and entertain other people. I think it ridiculous for anyone to listen to Casey Kasem when there are Cooper's hawks to watch. However, we recreationists—good and bad—seldom interfere with each other or even have to meet, which is one of the excellent things about rivers in general and these Ozark streams in particular.

If I had the keys to this place for a day or two, I wouldn't change a thing but its name. I would make it the Ozark National Scenic and Messing-Around Riverways. This would more accurately describe how things are there and might even make it a bit easier for the Park Service to fulfill its mission.

Reflections on Hawking

Concerning Jargon

A falconer is someone who gets hold of and trains (the process is called manning) a raptorial bird so that it will fly freely, hunt suitable prey, and return on command to the person who has manned it.

During the three or four millenniums since its invention, a lot of species of birds of prey have been used for falconry. Currently in North America those most commonly employed are the peregrine; gyrfalcon; merlin; kestrel; prairie falcon; goshawk; Cooper's, red-tailed, Harris', and rough-legged hawks; golden eagle; and great horned owl. Anyone who has manned one of these birds can be called a falconer.

Taxonomically, falcon refers to members of species belonging to the family Falconidae. Therefore sticklers for form insist that only those who man one of these long-winged predators are entitled to call themselves falconers. Then there are the exquisite purists who contend that only the female peregrine, long regarded as the *crème de la crème* of the birds of falconry, should be called a falcon. The male is a tercel.

There is another, more homely, parallel system of nomenclature wherein any manned bird, regardless of species or sex, is called a hawk, those who man them hawkers, and the entire activity hawking. These terms are so little favored by contemporary falconers that to use them is a bit of an affectation. However, I like to employ them because they tend to annoy and deflate hard-core falconers. This is a good work, because these people, if not checked now and then, often start talking, acting, striking poses, and generally prancing around as if they were characters out of bad Sir Walter Scott novels.

The Banana Box File

Beside me is a cardboard box, originally intended for the shipping of bananas, which is packed with copies of taped telephone conversations, court transcripts, editorials, news stories, private and public letters, all having to do with falconry and the law. Its contents weigh twenty-two pounds, seven ounces. This collection was begun early in the summer of 1984 when law enforcement agents of the U.S. Fish and Wildlife Service, after a three-year undercover operation, busted a bunch of hawkers—and/or traffickers in hawks—who federal agents claimed were doing illegal things with birds of prey.

The announcements drew an immediate and hostile response from officials of the North American Falconers Association, whose members account for a little more than half of the estimated twenty-five hundred hawkers in this country. Falconer spokesmen ever since have been claiming that the federal bust was conducted in a high-handed, despotic, if not illegal way; naive falconers were enticed and entrapped into criminality; and the scope of the operation was puffed up to make the federal cops look good for—among other reasons—budget purposes. The people actually arrested have not had much to say. In forty-two cases thus far settled, forty of those charged have pleaded or been found guilty, and two men have been acquitted.

The agents of course defended themselves, suggesting that they have done a big, good law enforcement thing. Colleagues in the Department of the Interior and allies in private conservation organizations are inclined to be suspicious of falconry and came to the support of the agents. There is every prospect that the controversy and accretion of paper will continue. In the future the federal agents intend to bring charges against approximately forty more falconers, Don Carr, the Justice Department attorney in charge, told me.

Some of the materials in the banana box were sent to me unsolicited by parties to the dispute who have axes to grind. They figured that since my vocation is commenting in public prints on wildlife and general environmental matters, I would get around to writing about the Great Falconry Bust of 1984 and, if given the right information, might grind their axes. The rest of the documents I have sought out myself. First, I suspected that indeed I would feel obliged to deliver myself of a few thousand words on the subject. This has obviously come to pass. Second, I was once an obsessed falconer. This should be entered on the record early, to warn against biases on this subject, of which, admittedly, I have many.

A second caveat is in order. Anyone craving a serious analysis of the falconry sting operation will be disappointed in what follows and should apply elsewhere. This happening has received a great deal of media attention, principally, I think, because most things having to do with falconry strike a lot of people who don't know a hawk from a handsaw as being esoteric in the same way that accounts of witches, snakes, and cattle mutilations are. Throw in some felonies and rich Arabs and you have a story with a high titillation content. However, what the Fish and Wildlife Service has rather grandly called Operation Falcon seems to me to be no more important than any other run-of-the-mill anti-poaching exercise. Some people broke the law, got nabbed, and their associates are mad about it, making excuses for them.

Why some otherwise upstanding citizens, teachers, physicians, and entrepreneurs so coveted a hawk that they risked paying stiff fines, even doing time in federal slammers to get one — not how they got caught — is, in my opinion, why Operation Falcon is worth writing about. What follows has to do with the nature of falconry, with emphasis on its metaphysical aspects — and some speculation about whether or not it is compatible with contemporary society.

Remembrances of Hawkers Past

Falconry was once what jogging and NFL football are now to us. This may be a suggestive analogy, but it is an inadequate one. Running about in our underwear or watching fat men push each other around plastic-covered fields may prove to be only a passing fad and is no more than a regional one. Falconry massively engaged much of the civilized world, from the English Channel to the Sea of Japan, for a millennium.

Marco Polo reported that Kublai Khan employed ten thousand professional falconers to get birds for his mews—the places hawks are kept—and manage them. Like the proprietor of a major harem, Kublai had his favorites, but a squad of executive falconers was always in attendance, ready to present a new bird for his pleasure.

At about the same time on the other side of the world, Juliana Barnes, an English abbess, wrote a treatise on the social and legal ramifications of falconry. She tells us that by custom and law only emperors were held to be classy enough to carry and fly gyrfalcons, the largest and rarest falcon then known in western Europe. Peregrines were suitable for kings, dukes, and barons of exceptional clout. Merlins were for ladies and goshawks (pothawks) for yeomen. The last people Juliana thought worth mentioning were holy-water clerks, who were entitled to no more than a musket—the male of the European sparrow hawk.

However, there were scofflaws, people who kept hawks that were too good for them. This was regarded as an act of defiance against the established social order and also as a potential threat to a valuable wildlife resource, making good birds harder and more expensive to get for those persons whom a Good Lord obviously intended to be their consumers. The Abbess Juliana wrote that cutting off the hands of people who got above themselves hawkwise had proved to be an excellent deterrent to this sort of felony.

King James I wrote in the sixteenth century that falconry was "an extreme stirrer up of passions." The observation may be even truer now, since the activity has declined in status and scope to that of a small, peculiar sect. The traditional passions of contemporary falconers often are reinforced with a strong streak of paranoia. Like gypsies, fruitarians, pious rattlesnake handlers, and other intense cultists, they tend to see the rest of the world as ignorant of and hostile toward their interests.

The phrase "obsessed falconer" is, these days, redundant, as describing somebody as a bigoted racist or agile gymnast would be. There may be among them the equivalent of a weekend golfer or occasional bridge player, but I have never met or heard about any. To go further, I have been around a lot of hard-core whitewater canoeists, cavers, rock climbers, bridge and golf players, who get very steamed up about these pursuits. However, in my experience, falconry can hook vulnerable personalities harder and deeper than any other ostensibly avocational activity.

All falconers I have known can be fairly described as being obsessive, but they come in two classes. Normal falconers devote most of their free time to their birds and are forever trying to enlarge this part of their life by stealing time, like sly kleptomaniacs, from other areas. However, they are able to function in the real world. They work and discharge minimal domestic and civic obligations, though while doing so at least part of their thoughts are on hawking. Quantitatively Kublai Khan was perhaps the grandest of all falconers, but he seems to have been a normal one, since he also ran a successful empire.

Bill Harry is another, living, normal falconer. People have told me that he is also a respected engineer-researcher, important in the military-industrial complex, but I don't know much about this part of his life. In our youth we spent a lot of enjoyable time together catching hawks and sharing intimate thoughts and feelings about hawking. We retain nostalgic feelings about each other, but in the last ten years have drifted apart because I drifted out of falconry.

I called Bill a few months ago and asked if he had been indicted by the feds or had his phone tapped. He said he thought he was clean because he no longer had a bird. His last peregrine, which he had flown stylishly and skillfully for many years, had died the previous winter. What with children in college, changes in his professional schedule, and other things, he could not see his way clear to keeping, properly, another bird. "You know the saying . . . there can be life after falconry. I guess it's true. But, damn, it's a hard way to live," Bill commented.

The second class of falconers is made up of those whose passion has consumed them, become their life—those who by ordinary behavioral standards have been driven bats by it. No names will be mentioned here because some of them are dead, others may have reformed, and, if they have not, it seems cruel to identify them. But there have been hawkers who have left jobs, marriages, families, love affairs, society (except for that of a few fellow addicts) because these arrangements and relationships interfered with falconry. There was a fellow who, after taking a young gyrfalcon from an Arctic eyrie, became lost in the wilderness for a week or so. Both survived, but the bird was in better shape than the man, because the latter had fed the former with pieces of himself. There was an inseparable pair of falconers who were, in a nonsexual way, true lovers. They got into a dispute about the ownership of a goshawk and never thereafter met or spoke but outrageously slandered each other. There was another pair of hawking mates who trapped peregrines on one of the Atlantic beaches. One day they lost a beautiful bird from their nets, fell to arguing about whose fault it was, and then brutally beat each other. The next day they went back to trapping together and remained close friends until one of them died.

These days members of the falconry establishment do not like hearing such stories. I have been told that even if my information is accurate, it is obsolete and has to do with happenings and people from the bad old days of fifteen years or more ago. True falconers now are responsible hobbyists like basement woodworkers or serious aviculturists, students of the taxonomy and behavior of raptors. This party line is understandable—what with all the bad publicity from Operation Falcon—but I do not find it entirely convincing that after so many centuries of passion, falconry suddenly became, circa 1970, a sedate branch of poultry-keeping.

However, it is true that I am no longer well connected in hawking circles. Therefore I was glad to come across a book published in 1984 by Stephen Bodio, a young man who is a very *au courant,* establishment falconer. Bodio writes mainly about up-to-date techniques of manning and flying birds, but he is too good and honest a storyteller not to make mention of the temperament of contemporary hawkers, which seems about like that of those I knew. In one anecdote, an impassioned young hawker complains bitterly about a proposed public law that threatens to restrict his freedom to hawk, saying, "Man, they can't do that. I mean it's like messing with my religion." The title of Bodio's book, *A Rage for Falcons,* is in itself instructive and, to me, reassuring.

Maybe this is what I am trying to say. Physically, falconry is about as risky as carpentry; the worst thing that can happen is having a bird drive a talon through your hand. However, psychically it is one of the ultimate thrill sports. In it there is a sense of standing very near the edge of a great, bottomless pit, the Abyss of Monomania. The elemental thrill and addiction of falconry may come from disporting oneself on the brink of the abyss. Normal falconers stay a safe half-step or so from the edge. The others by accident or design go off the deep end.

How Hawking Is Caught

I think that because of early environmental experiences, or perhaps heredity, there are certain personalities especially susceptible to falconry, just as others are predisposed to obesity, astrology, or Kahlil Gibran. First, these types have, for one reason or another, what C. S. Lewis has identified as a longing to know other bloods, i.e., an overriding sense that seeking ever more intimate connection with other species is an exquisite pleasure and important—spiritually and intellectually. Second, they are true romantics. The word is used as conventionally defined. Romantics trust and are moved more by intuition and imagination than by reason and experience. Romantics are inclined toward individualistic expression and action, are against authority, convention, classical modes of thought and behavior: young Werther, Lord Byron, Thoreau, James Dean.

Romantics who are keen on other bloods generally discover and come to admire birds of prey extravagantly in their formative years because these creatures are the pluperfect symbols of the free and bold. They adorn our flags and battle standards, seals and coins, tokens and totems. The notion that raptors are the ultimate representatives of the wild and romantic has been trickling into the human imagination for so long that it has become an indelible component of it.

Romanticism and animalism do not lead directly or inevitably to falconry. These traits have produced a wide, gaudy range of other obsessions: from writing poetry for and about cats to hunting tigers. However, in modern times there probably has not been a falconer who was not also a romantic animalist.

Another extremely common but perhaps not definitive characteristic of the falconer personality is maleness. It is true that Lancelot and Guinevere courted while hawking, but a close study of surviving records indicates that she had a bird because she was trying to be his bird.

It was a way to show she was a good supportive sport. In my own time I have known some fine though probably misguided women who have put up with a lot of hawking and developed some interest in various aspects of it, but I have the strong feeling that they did so for the sake of addicted men, not for hawks.

Other records are indicative. The bibliography of falconry works is now a huge one, having been accumulating for more than a thousand years. In it is only one substantive entry by a woman, the Abbess Juliana, and hers was in reality an etiquette not a falconry book. Last year all the people caught in the great federal falcon bust were men.

Through the centuries it has been implied or plainly stated that women do not have enough mind, heart, or stomach for falconry and therefore it should remain a macho activity. All of this is a cover story for a taboo that has applied to a great many things other than falconry for a long time. It is that women should be passionate—abandon themselves to be carried where their senses take them—but only in circumstances, usually sexual, deemed natural, moral, and legal by men. This ancient, widespread notion came to be accepted by both sexes and in consequence women who abandoned themselves to pursuits unrelated to and unsupervised by men have been generally regarded as unnatural creatures.

I am told by veracious people that there are now some women who are obsessed by falconry for its own sake. This is a big but not surprising change, since in the last decade or so the whistling-girls-and-crowing-hens-always-come-to-some-bad-end shibboleth has come under effective attack. Probably we now have some passionate female falconers for the same reasons that there now are more passionate female scientists, artists, and politicians.

Animal-loving, romantic boys (to deal with things as they still mostly are) sometimes become self-taught or infected falconers. They can be set off by very small sparks—seeing too much of Prince Valiant or Kirk Douglas, reading an "Ancient Sport Still Flourishes" bit of journalism. A long time ago I did one or two of these but stopped, in part because the vision of youths twitching and drooling over the words made me feel like a zoological pornographer.

More commonly, those of susceptible nature meet a practicing falconer and catch it from him. In medieval days this happened routinely, since the management of a hawk was considered as couth an attainment as handling a sword or battle-ax. Now the traditions and skills are transmitted largely through serendipitously formed relationships between potential and practicing hawkers.

My Days as a Hawker

At least in its early stages my case history is ordinary. Largely because of the influence of a naturalist father, I came very early to like all animals a lot and especially birds of prey. I was a moony child, a great consumer and creator of romances. But I grew up somewhat, and after World War II I settled in the Washington, D.C., vicinity. There I inevitably became acquainted with some falconers because this area has been a hotbed of hawking for the past fifty years. One of them who was leaving the area gave me my first bird, a young Cooper's hawk, which I soon lost because it was imperfectly manned. After that there was another Cooper's, followed by a goshawk, a peregrine, and a lot of redtails, which I found was the species I liked best. I once gave a fine, mature, wild-caught peregrine to another falconer because keeping it would have interfered with keeping redtails—and this marked me forever as unsound among orthodox cultists.

Though aesthetically spectacular, peregrines always struck me as a bit shallow of personality, specialized creatures like beauty queens or quarterbacks. Also, they are not well suited for flying or being retrieved in heavily wooded country such as central Pennsylvania, where I lived for most of my time as a falconer. The accipiters are exciting birds but cantankerous, high-strung, and sometimes almost manic. I am glad I got to know some goshawks and Cooper's hawks, but they are as exasperating as pit bulldogs to live with for any extended period of time.

Redtails, perhaps because they are by nature generalists, seem to me to be steadier, more sensible and interesting birds. They are big, burly, hard-working hunters that will adapt to all sorts of terrain and situations. If, for example, one misses a rabbit in an open field, she will, if necessary, go into a blackberry thicket on foot after it.

Toward the end of my hawking days, taking game with a redtail became an incidental interest. I would carry the bird on my fist to an overgrown field or woodlot and let her fly. As I strolled along she would follow, generally taking short hops between trees or bushes, sometimes rising above the woods to get a better view, occasionally finding something to pursue, much less often catching it. When it was time to go, I would whistle her back to my fist. Simply walking with and watching the bird became for me the great pleasure of falconry. My habit was to take a redtail in the late fall, get her ready to fly before Christmas, and then walk two or three times a week with her during the winter months. In March I'd set her free in an isolated area, to go back to doing whatever she had been doing before we got together.

The basic requirement for any sort of hawking is to have a hawk. In the circle where I was instructed, taking eyas or brancher birds (ones in or around a nest) was considered a poor, amateurish method of procurement. Raising such young birds properly is tedious, and they are never quite so expert or interesting hunters as birds that have acquired their skills in the wild. Even in those days birds could occasionally—no doubt illegally—be bought, but this was thought to be as wimpy as paying for sex. The decent, manly thing to do was to go out during the fall migration and take a fully developed bird.

There are various ways of catching a hawk, but I think the most satisfying is to use a bow net. This involves setting up a blind where hawks are concentrated—along a migration route—using live pigeons to lure them into the circumference of a spring-loaded circular net which is triggered by hand. Bow-netting has been described as fly-fishing in the sky, but I think this is inadequate. Dealing with two pigeons, three lines, and a hawk that may be diving at seventy-five miles per hour is, in terms of skill and stimulation, beyond compare.

Early on I got a federal permit to take and band birds of prey, and this allowed me to bow-net with some legality. I would keep a bird a year for hawking, but the others—and there were some hundreds—were taken from the nets, measured, sexed, banded, and released. Theoretically this is of some scientific value, but when I spent thirty or forty days each fall sitting on cold Appalachian ridges, angling to bring down birds to the net, it was for the pure joy of it. From the beginning it was as exciting as hawking itself, and after a time it became more so.

The hawking passion had the worst hold on me in 1955. I am certain about the time because our first daughter was born on December 7 of that year. Throughout the fall I was mainly, in body and spirit, with the birds. My wife, Ann, went a good many extra miles being patient and sympathetic about my affliction, but around Thanksgiving she made such an accurate and ironic observation on my priorities that it got through. She said, "Maybe we should name the baby Redtail."

In fact we named her Lyn. Later she had a barred owl as a playpen mate, a kestrel of her own, and sometimes went bow-netting. But by then things had changed. I am still romantic enough and tend to be easily caught up in passions, but I also am a generalist, like a redtail. While hawking I nurtured passions to explore the Arctic barren lands and coral reefs of the South Seas, but I also wanted to be a writer and—to summarize a complex of personal desires and hopes—have a daughter named Lyn rather than one named Redtail. In any event I came to accept the logic of the Irish aphorism "A man cannot be in two places at once unless he is a bird."

For the next twenty years I spent time each fall with bow nets but sometimes did not keep a bird. I last kept a redtail fifteen years ago. Eventually I stopped altogether because—though sometimes I missed the bow-netting painfully—other interests took me elsewhere. This past fall, for the first time in five years, I stayed in the Appalachians to go after hawks. But at the old trapping site, more often than not, I didn't bother to set the net. Sitting on the rocks, in the thin air, mingling with the migrants was plenty. It seems I have come full circle and am back where I started fifty years ago as a hawk appreciator.

I am not a falconer anymore, and in a way I am glad. However, this is not to say I am sorry I once was one. There is within me a seam that was laid down by hawking. Since on balance I would rather be what I am than something else, I am grateful for the experience.

If I were to know the time of my final departure from these parts, I think I would like to spend an hour or so on the day before walking in the Pennsylvania brush with a redtail, or at least be able to clearly remember those times—for they were ones when I thought I had a good sense of this place and liked it very much.

I am indebted to hawking for other, more abstract reasons. Falconers often give the impression that manning a bird is a feat comparable to learning Sanskrit or manufacturing a nuclear device. This is puffery. Cultists, like medieval monks arguing about the number of angels that can cavort on a pin, will carry on interminably about fine technical points, but training a bird essentially involves only two rather simple things. The man must fuss with the animal enough that its fear is dissipated. Second, he must condition the bird to accept food from him and come to him for it. A wild-caught hawk can often be flown—and will return to the fist or lure—within six weeks after its capture. Training a good sheep dog or cow pony is physically and intellectually a much larger accomplishment.

But manning a hawk is an excellent exercise for teaching patience and self-control. Unlike dogs, horses, or other mammals whose cognitive powers are more like our own, birds of prey cannot be forced to do anything except flee or defend themselves. You could beat a hawk to a pulp and not come a bit closer to getting it to do what you want. At least among themselves, falconers admit that the desire to take one of these pea-brained, mortally stubborn creatures and wring its aggravating neck sometimes rises like bile in the throat. On the other hand, I have never heard of a falconer who has abused a hawk.

At least in my case, trying to deal with a hawk has substantially influenced my opinion about the proper way of dealing with mountains, rivers, some species of flora and fauna, and other more complex inhuman phenomena. The raptors belong to a large class of natural things which we can break and destroy but which we cannot coerce into being useful or enjoyable. We must accept and adjust to their alien essence rather than trying to change them into something else. Manning a hawk is an effective antidote for hubris, which in my opinion is a very serious affliction today, one which is particularly dangerous because it seems to attack the faculties that enable us to identify and adjust to reality.

Hawking was good to me in terms of pleasure and instruction. And as a bonus I developed a healthy fear of monomania. Ever since, I have been mistrustful of zealots and fanatics of all sorts because the first thing that seems to burn out among those in chronic and passionate pursuit of the ideal is a sense of humanity—theirs as well as everybody else's. The consequences are trivial if only hawks are involved, but often serious when this sort of idealism is focused on larger pursuits—say, manning people.

Hawking, the Law, and Public Opinion

Until about fifteen years ago, falconry existed in a legal gray area, being neither prohibited nor permitted in most jurisdictions. In consequence, falconers were chronically insecure, always a little fearful that a cop would pop out of the bushes and say, "Hey, buddy, what are you doing with that buzzard?"

Falconers were—and it would seem still are—even more nervous about public opinion. The assumption is that falconry has a mysterious but generally wicked reputation; that a bit of bad publicity could lead to banning the whole activity, and that this could easily be done since falconers are a small, odd minority. These suspicions are not entirely paranoid. In a large segment of the public—people interested in wildlife affairs but not acquainted with falconry—there is indeed a common notion that it is a cruel, violent, destructive activity; that it is roughly comparable to cock- or dog-fighting and deserves to be treated as such. This opinion is, I think, based on a specific ignorance of falconry and indicates a general prejudice against the consumptive use of wildlife.

In regard to ignorance: The vision of hawkers and their birds committing predacious orgies against rabbits and ducks is a fantasy. A manned hawk is vastly inferior to a gun, snare, or even a ferret as a game-getting device. When a bird is flown at prey it often misses. But either way, when the bird is retrieved it must be fed, and the more it is fed the more unresponsive it becomes. If a hawker gets greedy about hunting, he will lose his bird, and he would rather be a vegetarian than have that happen.

There is some blood in falconry, but it mostly is the blood of the pigeons used to catch, train, and feed hawks. And in fact the part of hawking that most hawkers like least is the sacrifice of pigeons— animals of excellent and attractive character who are in many ways more interesting than birds of prey. In consequence, falconers often become genuine pigeon fanciers.

A hawk is the most prized possession most hawkers have. They dote on their birds and worry constantly about their physical and psychic well-being. A truly detached observer, say, an ethnologist from Mars, observing falconry would probably conclude that he had come upon an odd slave–master arrangement in which the bird filled the latter and the man the former role.

An important corollary criticism is that falconry is cruel because it involves holding a wild thing captive. These feelings are understandable—particularly because raptors are such powerful symbols of freedom—but I suspect they are based more on anthropomorphism than on fact. There is little evidence that the abstract concept of liberty much engages other bloods or—if they are well fed, housed, and entertained—that the lack of it creates great anguish among them. I have spent a lot of time watching wild hawks, and my impression is that when they are not hungry or otherwise uncomfortable they choose to do about what a manned one does, sit on a perch and turn off its mind. Furthermore, of all the animals we catch and keep, a hawk probably undergoes the least alteration of its life-style, being able to fly and hunt frequently. If captive animals do pine for their freedom, the problem probably is much more prevalent in the carnivore and monkey houses of zoos and among dogs chained in backyards than it is in the mews of falconers.

When I was a falconer, there were two theories about how to avoid getting crosswise with the law or public opinion. The first was to keep a low profile and work with one's bird without having to fill out a lot of paper and permits. The second—favored by those in falconry clubs—was to seek recognition, trading some legal regulation for some legal protection. In the 1960s these organized falconers began lobbying, quite effectively, in state legislatures and game departments. Now falconry has some legal status in most states.

As for federal laws, the Lacey Act, which regulates trafficking in wildlife, has been in force for most of this century and gives federal agents some authority to deal with falconers. But for a long time they seldom did, presumably because the traffic seemed so insignificant. This changed with the passage of the Endangered Species Act of 1973 and its application to the American peregrine. This species presented a special case, since a lot of peregrines were already in captivity. For several years both the agents and falconers worried about the issue. Then in 1978 the act was amended with a kind of grandfather clause stating that anyone who possessed a peregrine prior to 1978 could register it and keep it, but thereafter any taking of wild peregrines would be treated as a felony. Some poaching of peregrines continues, but nobody is certain how many birds are involved. Twenty-five to fifty a year is a frequent guess.

Having become interested, the Fish and Wildlife Service turned its attention to all birds used for falconry and now has guidelines which set up a system similar to that described by the Abbess Juliana Barnes in the thirteenth century, though the penalties are less harsh. The guidelines divided falconers into three classes—master, intermediate, and apprentice—and then specified which raptors were suitable for members of each class: redtails and great horned owls for apprentices; the large, true falcons for masters; and in-between birds for intermediate people. Many states have adopted these guidelines as they stand but have the option of making them even more restrictive.

There is now another complicated regulatory matter involving falconry. During the past fifteen years great advances have been made in breeding birds of prey in captivity. According to a report made by Bob Berry, president of the North American Raptor Breeders Association, there are about two hundred legal propagators whose twelve hundred captives produced some six hundred offspring in 1983 alone. Most of these birds are peregrine subspecies, gyrfalcons, and Harris' hawks, but now techniques are available for raising all of the birds used in falconry. The products of these efforts, including some hybrids, are being called "domestic quasi-wildlife." Enthusiasts claim that henceforth improved models of birds will be produced like hatchery pheasants or poodles in kennels. Already raptor stud books are in existence.

State and federal authorities as well as wildlife conservationists have been fairly suspicious of these enterprises because they are new and because falconers have the reputation of being so obsessed with getting birds. In consequence, captive breeders must permit law enforcement agents to inspect their premises without warrants and must make frequent and voluminous reports about the birds they have, their reproductive and mortality records. Also, a captive-raised bird must wear a seamless band devised and supplied by the federal authorities. This can be affixed only before the bird is two weeks old and in theory cannot thereafter be removed without showing signs of tampering. The fear in all of this is that falconers will use captive breeding as a cover for obtaining wild birds, especially peregrines, and passing them off as domestic quasi-wildlife.

Initially, captive-bred raptors could not be bought or sold, but breeders complained that rearing the birds was a time-consuming, fairly expensive operation and that they should be able to at least break even. They argued that making captive-bred birds available would relieve the pressure on wild ones—again the peregrine is the critical species. This argument was accepted by the Fish and Wildlife Service, which in 1983 issued regulations that permitted commerce in birds by licensed breeders.

Several conservation organizations—including the National Audubon Society—objected to the change on the grounds that if people started profiting from raptors, there would be even more incentive for taking wild ones and "laundering" them by switching seamless bands. (These bands, it seems, are not entirely tamperproof.) The first announcements concerning Operation Falcon strongly implied that some of those arrested had done just this. However, later information indicated that all of the cases involved offenses, or allegations thereof, which had occurred before the passage of the regulation that permits commercial dealings in raptors. Therefore the question of whether or not trading in domestic quasi-wildlife has a good or bad effect on real wildlife remains to be answered. However, it has clearly created the opportunity for a lot of new regulatory works.

Operation Falcon

Rick Leach is the assistant chief of law enforcement for the Fish and Wildlife Service and has been in charge of Operation Falcon since it was launched in 1981. When we talked last fall I asked him how much the exercise had cost in terms of manpower and money, and he said that in his shop they did not answer that sort of question. However, a press release issued by the Fish and Wildlife Service notes that at certain times one hundred fifty agents—there are two hundred all told—were involved. No mention has been made of money, but as an educated guess about a million dollars of public funds probably will be spent on this work.

As shown in various legal documents, the principal undercover operative was one Jeffrey McPartlin. Now forty, he has been a falconer since his early teens and, it would appear, a monomaniacal one, at least until he got into the informant business. I asked Rick Leach if McPartlin, still a federal employee at $2,200 a month, was granting interviews. Leach said he was not. I probably could have forced the matter but did not because I saw no good reason for making new acquaintances of this sort. Also, a lot of biographical material about McPartlin is now available. The North American Falconers Association is a good, if understandably hostile, source.

While moving about the West, always engaged in falconry, McPartlin was convicted in the early 1970s of illegally trafficking in raptors. This record, and a reputation as one who sailed very close to the wind as far as falconry laws were concerned, recommended him to the Fish and Wildlife Service as a sting man. To set McPartlin up in the undercover business, federal agents trapped forty-two wild raptors: thirty-six gyrfalcons, three prairie falcons and three goshawks. Some of these were later traded for seven already captive peregrines. McPartlin was also involved in an expedition during which a falconer poached three peregrine eggs. Falconers contend that McPartlin arranged and led this caper, incubated the contraband eggs, and therefore was guilty of entrapment and a violation of the Endangered Species Act, since not even on-duty federal agents may molest creatures protected by this law. The agents say McPartlin was only an observer of wrongdoing, not the perpetrator of it.

So, supplied legally with birds for illegal sale, McPartlin began, in a manner of speaking, hawking his wares. According to court records he told some Texas physicians he had a few peregrines that were "nice, big, black, and beautiful as hell." He then jocularly reminded the customers that "the illegal ones fly better than the legal ones." The falconry community has charged that McPartlin entrapped some of its members. Thus far the courts have ruled otherwise, saying that the sting man did no more than offer people "the opportunity to violate the law," which is acceptable in these operations.

On the basis largely of evidence collected by McPartlin, one hundred fifty federal agents, on June 29, 1984, swooped down on thirty-two falconers in fourteen states and arrested them for various alleged offenses. Simultaneously, the Fish and Wildlife Service issued a press release describing the bust. Among other things, it said that the agents had uncovered a "multi-million-dollar" black-market operation that involved four hundred birds taken from the wild.

After having had a chance to study the announcement, falconry spokesmen replied that the "multi-million-dollar" claim was a gross exaggeration, based on evidence which showed $150,000 at most changing hands. The largest single transaction was between McPartlin and a West German, Lothar Ciesielski, in which the latter paid $112,000 for nineteen gyrfalcons. Ciesielski is a member of a family that deals extensively in raptors, and this batch was allegedly to be resold to Saudi Arabian royalty.

Furthermore, falconers critical of the operation claimed the evidence presented showed that only fifty-three birds—not four hundred—in addition to the forty-two procured by federal agents for McPartlin had been taken from the wild. Finally it was argued that the only dealing had been done by the Ciesielskis, some Canadian hustlers, and McPartlin. The American falconers arrested had simply wanted to get birds for themselves, had obtained them from McPartlin, or had got them elsewhere. In the process they had violated the Lacey Act in regard to interstate traffic or had falsified records or not kept accurate ones of how they got their birds. The falconers contended that these were small, technical offenses that did not warrant the agents' carrying on as if they had taken out a Mafia family.

Speaking for the federal agents, both Don Carr of the Justice Department and Rick Leach of the Fish and Wildlife Service said that, while many of those in the first round of arrests had not actually committed major offenses, these cases are leading up to much larger ones which will be announced. Both Carr and Leach said that when all the evidence is made public, the original figures about money and birds involved will be justified.

It seems to me that the federal announcements were indeed a bit strident and excessive, though not so much malicious as an exercise in ordinary bureaucratic chest-pounding. In some cases the federal releases were misleading—the inference that commercial raptor breeders were "laundering" birds—and in others premature, to say the least. For example, from the accompanying evidence it would have been more accurate to have announced, "Federal agents spend a million to uncover $150,000 falconry black market."

The bottom line of Operation Falcon is that, in all but two of the cases thus far resolved, falconers, because of their obsessive feeling about birds, knowingly violated the laws of the land. McPartlin may have been a bit sly, but he did not force them to do what they did. To claim, after the fact, that the violations were not major ones and that the laws themselves may be Mickey Mouse does not cut any ice legally or logically.

So What?

Operation Falcon is of more than passing interest only because it has focused attention on a question that has hovered over falconry for some time. Is it in the public interest to sanction this activity?

Obviously if falconry injured or seriously discommoded other people, there would be good reason to prohibit it. It is my contention that hawking can have a bad effect on the mental health of its devotees, but they tend to keep to themselves, do not employ dangerous weapons, make loud distracting noises, or despoil the property of others. In short, there is no evidence that falconers, as such, create general public health or nuisance problems.

This being the case, the crucial issue is whether falconry is bad for publicly owned wildlife, specifically for raptorial birds. I took this question to four persons who have exceptional knowledge about the subject. They are: Jim Ruos, a migratory bird biologist with the Fish and Wildlife Service and also a veteran, expert falconer; two of his federal colleagues, John Spinks and Dan James, biologists with the Endangered Species office; and Brian Millsap, a raptor biologist for the Raptor Information Center of the National Wildlife Federation and an ex-falconer.

The four men had virtually identical answers to the question. It was their opinion that in addition to habitat reduction and degradation—obviously a problem not confined to raptors—there are two areas in which human activities have had significant adverse effect on birds of prey. At one time thousands were killed annually by gunners, either for sport or because they thought they were protecting game species from these predators. Also, the widespread use of DDT had a terrible effect, especially on the reproduction of most of the raptors. Now that shooting and the use of DDT have been banned, all of the species used in falconry except the peregrine have made a good recovery.

As to other activities that cause mortality among raptors, the four biologists named three of about equal—and minor—importance. They are: birds injuring or killing themselves by flying into buildings, fences, or vehicles; accidental shootings by sport hunters, mostly waterfowlers; and the taking of hawks by and for falconers. All felt that as presently practiced falconry did not pose any threat to *most* of the species. Brian Millsap suggested that in certain restricted local areas, populations of Harris' and Cooper's hawks might be temporarily depleted because they were being taken by hawkers. As to the peregrine, all agreed that some wild birds were poached for falconry but said it is not known if this significantly contributed to reducing peregrines to endangered status or is inhibiting the recovery of the species. However, they were unanimous in suggesting that because of the scarcity of peregrines, the removal of any of them should be treated as a serious matter until it is clearly demonstrated not to be.

As far as I know, there are no other professional opinions that differ markedly from these. Since falconry creates at most very small problems in the area of wildlife protection, it might be in the general interest if, officially, we worried about it less than is now the case. The complex of law and bureaucracy now in place to supervise the practice of hawking seems excessive. It has evolved not as a solution to a critical problem but in response to notions that this activity is contrary to modern environmental sensibilities and therefore requires a lot of watching. I think it can be argued that at present the most objectionable thing about falconry is its over-regulation, which, because of the fantasies about falconry, distracts public attention and diverts public money and civil servants from more important matters—say, increasing protection for wild peregrines.

A good place to start deregulating hawking would be to scrap the current guidelines that classify and match up hawks and hawkers. This seems as silly as it would be for federal agents to get into the business of handicapping golfers. The present rules could be replaced by a simple statement listing the birds of prey which in the federal opinion are numerous enough that some of them can be harvested by falconers. It would then be up to the states—those that wanted falconers—to accept this list or modify it to protect additional species. A state falconry permit would be issued just as permits are given to hunters. The fee should be high enough to cover any extra law enforcement expenses, and it might also be well to prohibit the harvesting or keeping of more than one wild raptor at a time. Anyone receiving a permit would be entitled to have any approved raptor as long as ordinary hunting laws or those having to do with inhumane treatment of animals were not violated. These probably would still be some violators, but they should be no more difficult for wardens to catch than deer or duck poachers.

I can see no way in which the public interest would be served by including the peregrine among the birds suitable for falconry, at least until the species fully recovers. As to those already in captivity, we should be able to indelibly mark these birds, perhaps with tattoos. Falconers would be given a few months to identify their peregrines, and thereafter any birds not so marked would be illegal. After the demise of the last of the registered captives, no peregrines, regardless of origin, could be used for falconry. Hawkers would carry on piteously that such a prohibition interfered with their civil and religious rights, but no attention should be paid to them. The substance of falconry—the passion, pleasure, and instruction—can be had from redtails and other common species.

As to captive propagation, my feeling is that in time there will be two activities based on raptors. Falconers will go on working with birds from the wild, while high-tech hobbyists will play with domestic quasi-wildlife. However, since captive propagation has been invented, it, like nuclear weapons, probably cannot be disinvented and therefore needs some regulation—but again less than is now the case. Breeding peregrines in captivity should be phased out to avoid any hanky-panky—obsessed falconers' augmenting their stocks with birds from the wild. Otherwise, propagators working with approved species, which they were able to prove were of domestic origin, should be allowed to do so as freely as people raise fancy chickens. Whether or not such business would be profitable without the peregrine is of no public concern.

Getting Close to Wildlife

I think it can be fairly argued that, as it stands, falconry is virtually harmless and would be completely so if the use of the peregrine were prohibited and regulations modified so as to cause less public expense and commotion. Even so, there remains the question of why bother with it at all in any form. I think that, in addition to being innocuous, there are some positive factors which, in terms of the general welfare, recommend keeping falconers.

An obvious one was mentioned by all four raptor biologists I consulted: While the taking of birds, legally and illegally, constitutes a small drain on wild populations, the overall effect of falconry is preservative rather than destructive. This is because falconers constitute a small but aggressive pro-hawk constituency that supports protection of these species through habitat improvement, antipollution laws, and other environment-preservation programs. Falconers have also contributed labor, money, and expertise to organizations that rehabilitate injured raptors or that are trying to augment wild populations—particularly of peregrines—by the release of captive-propagated birds. This latter, well-publicized work is often brought up in self-congratulatory terms by falconers, and indeed it indicates a sense of social responsibility. However, some feel its practical importance may have been somewhat over-advertised.

As a professional bird biologist and falconer, Jim Ruos is involved in the release of captive-bred peregrines. He said of this effort: "It is challenging work and has given us a sense that we are doing something." Ruos expanded this faintly ironic comment by saying that in his opinion, barring the return of DDT, the commencement of nuclear hostilities, or other disasters, peregrines probably would be able to restore themselves in most of their original range without high-tech assistance from us.

There is a related and, to my mind, more important matter. In principle we are much more concerned about wildlife than we were fifty years ago. Generally this is all to the good, but one effect of this abstract interest has been to reduce the opportunity for truly intimate contact with wildlife. A child's raising an orphaned raccoon, fawn, or mockingbird—something common in my youth—is now, if not overtly illegal, regarded as an unenlightened, anti-environmental act.

Given all the arguments against such practices—that some of the creatures may be mistreated, that they could lead to mockingbirds' being sold in pet stores—I still think that if reasonably supervised, such intimate experiences are desirable. They benefit wildlife by expanding its genuine constituency. They benefit us by promoting in a visceral way the love of other bloods. I have never known anyone who had a pet crow who was not a good conservationist, and many of them became important scientific, political, and social defenders of wildlife. (I once argued, unsuccessfully of course, with John Gottschalk, a former director of the Fish and Wildlife Service, that the cause of conservation would be greatly advanced if a few of the national wildlife refuges, used to protect and produce ducks for gunners, were converted to crow factories where children could climb trees and get young birds to take home as companions.)

For these reasons and because the relationship between a hawker and hawk is one of the most intimate of all human-animal associations, I think falconry is desirable. In this connection, Jim Ruos told me about a colleague whose professional and private passion was for blue jays. The man confessed to Ruos that he had some resentment toward falconers because they had such special privileges. They could take and keep birds of prey while he was not permitted to do the same with blue jays simply for the pleasure of their company. I think the fellow has a legitimate complaint, but the solution is to make it easier and more respectable for him to enjoy blue jays rather than making it more difficult for others to enjoy hawks.

Finally, there are certain activities which in terms of conventional human behavior seem superficially to be only bizarre and extreme but which often provide a way to understand the extent of our possibilities. Among such things, at random, are samurai codes of honor, true romantic love, the righteousness of inquisitors, the visions mystics find in candle flames, and, I have always thought, falconry. Specifically, it dramatically demonstrates our susceptibility to pure, aesthetic and imaginative, passions which are quite distinct from material lusts. It is a good thing to keep falconry around because if even a few of us are capable of it, speculation must follow about what all of us are. Also, the temptation to make rash, pontifical judgments about what we can and should be is somewhat reduced. As a rule of thumb, tolerating more or less harmless eccentrics is socially expansive; rubbing them out advances nothing but the ideal of general conformity.

"God creates, Linnaeus arranges"

In western civilization, a list of the great culture heroes—Aquinas, Luther, Newton, Shakespeare, Michelangelo, Machiavelli, Napoleon, Lincoln—includes only two naturalists, Charles Darwin and Carl Linnaeus. At last estimate there were some ten thousand biographies, critiques, and commentaries on the lives and works of the pair, each accounting for approximately half the total. There is no way to make an absolute judgment about which is more renowned. However, Darwin's fame is easier to understand. Darwin thought and wrote in English, a much more accessible Western language than either Linnaeus' native Swedish or the Latin in which his works were published. Also, the mind-set of a nineteenth-century Englishman is closer to our own than that of an eighteenth-century Swede. Furthermore, Darwin's work—or at least some interpretations of it—is still controversial. While his insights about evolutionary forces have been accepted by the majority, there are refugee bands of cultural pagans on the lam from the twentieth century. Society's ills, they have determined, are largely the fault of somebody named Darwin. These peculiar but passionately held views have very little to do with the historical Charles Darwin or his theories, but they have served to keep the man's name current.

For a time—because he observed that plants have sexual organs and relations—Carl Linnaeus exercised the purity-prurient establishment of his day somewhat as Darwin still does. However, in two hundred years, floral hanky-panky has come to be widely accepted as a fact of nature, and neither this nor any of his other discoveries has the power to heat up politicians—or preachers. For that matter, though roughly everyone wtih a high school education or more has heard of Linnaeus, most will recall only that he had something to do with naming plants. Even the reasonably erudite are hard-pressed to explain why this was important, or what, if any, significance his activities still have. This is not particularly unusual, since time erodes the greatest reputations and Linnaeus died in 1778. However, what is very odd in the case of Linnaeus is that even the specialists with access to great quantities of information about the man have always had much the same difficulty—being specific and convincing about what made and has kept him so famous.

Describing and analyzing Linnaeus has become a fairly substantial cottage industry for scholars, and their biographical output is remarkably consistent in its judgment of the man and his activities. Linnaeus made no startling discoveries of natural facts, created no brilliant, comprehensive synthesis comparable to Darwin's theory of evolution. What he and his contemporaries regarded as his greatest work— establishing the Linnaean system for classifying animals, plants, and minerals—stood the test of time only briefly and badly. Key elements of it were rejected, within a generation after his death, as technically inadequate and philosophically wrongheaded.

His research and scholarship, in terms of objectivity and meticulousness—was often below that of a scientist of the first rank. As a personality there is abundant evidence that he also probably belongs a good many notches below first rank. He was a petty social climber, avaricious, desperate for honors and attention, and quite capable of fawning, dissembling, and simply cheating to advance his immediate self-interest. Above all else, it was the consensus of both his contemporaries and later biographers that he was a colossally, even pathologically, vain man. Among other unattractive activities, he published dozens of reviews and commentaries—both signed and anonymous—extravagantly praising himself and his work. One typical example:

In an autobiography—he was to write five—Linnaeus remarked, "God Himself guided him [habitually Linnaeus referred to himself in the third person] with His own almighty hand. He caused him to see more of creation than any other mortal before him. He bestowed on him the greatest insight into natural history, greater than any other had ever received."

With a certain air of desperation some Linnaean scholars contend that what he did, and who he was, is judged less important than when he did it. They say Linnaeus, despite his defects, deserves recognition as the first of the modern, enlightened naturalists. There is some truth but perhaps more irony in this theory. Linnaeus was a conservative creationist and a defender of the medieval view that the world was an eternally fixed and static place. His passion, personal and professional, was in finding and cataloging the wonders of nature. Linnaeus and most of his contemporaries saw this as a pious activity that glorified God, who had made nature just as it was and who would keep it that way as long as it pleased Him. In time the work and thought of Linnaeus led him and others in opposite directions, but he was a defender of the old ways rather than a prophet of the new or a theological rebel.

Even so, when all his flaws and inadequacies have been examined, there remains the incontestable fact of his continued renown. His name has been longer and more generally remembered than that of any post-Aristotelian naturalist. One possibility is that we have been victims of a long-standing hoax originated by Linnaeus himself, who, at least in the area of self-promotion, was a man of genius. Having listened to him proclaim his own greatness loud and long, his contemporaries may have been bamboozled into accepting it as fact.

A less cynical and more plausible explanation is that the continuing historical judgment is a true one—that Carl Linnaeus was one of those very rare birds whose specific parts, his accomplishments and ideas, seem no better than ordinary when examined individually but are extraordinary when considered collectively. At least this could partly account for the fact that we keep on adding up the man and puzzling over the great sum of him.

There are obviously many differences between our time and that of Linnaeus, but one of considerable significance is seldom noted. In Linnaeus' time, far fewer men—and almost no women—were able to support themselves by doing essentially cerebral work. Today the educational, sociological, health, communications, engineering, scientific, and bureaucratic industries employ millions of people who are, if not intellectuals, at least working with mind rather than muscle. In the eighteenth century and earlier these enterprises were either nonexistent or very small, primitive ones that offered few employment opportunities. The one substantial institution that utilized brainpower was the church—and from the sixteenth century onward the churches. Unquestionably many of the religious bureaucrats and academics were more inquisitorial than inquisitive, were unyielding zealots rather than free-ranging thinkers, but at least they were accustomed to using their minds. And not a few—Erasmus, Copernicus, Richelieu—let their minds wander in nontheological directions. By the seventeenth century natural history had become a common diversion for divines. They had the time to wander about, Gilbert White fashion, looking at birds, bugs, and blossoms, and they had the room in their parsonages to preserve and contemplate their collections.

This seemed like a theologically safe and uncontroversial pastime for a priest or parson, but in time that proved to be false. Eventually the rigid medieval theocracy was brought down, not by outside agitators and infidels but by restless, inquisitive churchmen and church-trained intellectuals. Surprisingly, the gentle naturalists were an important force in this revolution because they developed the habit of observing the world as it was rather than as ecclesiastical authority claimed it should be. This raised doubts and questions about larger philosophical matters. Ultimately it was one of these naturalists—the dreamy fossil hunter Darwin—who was to deliver the *coup de grâce* to sacerdotal science.

Carl Linnaeus was a hereditary member of the religious-intellectual class, born in 1707 in May, when, as he later wrote, "the cuckoo was announcing the imminence of summer, when the trees were in leaf but before the season of bloom." He grew up in the parsonage near Stenbrohult, a village in southern Sweden, where his father, Nils, served as a curate and then as pastor. His mother, Christine, was the daughter of a clergyman and granddaughter of two clergymen. Both his father and a paternal uncle—also a divine—were dedicated botanists, and a maternal great-grandmother was burned as a witch in 1622, when the line between science and sorcery was a very thin one.

Carl's father was a farmer's son, an Ingemarrson, but such were his botanical interests that when he went off to study for the ministry he changed his name to Linnaeus in commemoration of a great linden tree that grew on the family homestead. When he was established in the rectory at Stenbrohult he immediately laid out an exhibit garden, and he continued to develop it throughout his life, collecting interesting specimens as he made his pastoral rounds through the countryside. Almost as soon as he could toddle, Carl became a botanical follower of his father. "This as yet only son," Linnaeus was to write about his own childhood, "was, as it were, educated by his father in the garden. The latter designed . . . one of the finest gardens in the whole province, carefully selecting trees and rare flowers, and whenever he was free from the duties of his office spent his leisure time there."

It was perhaps inevitable that a boy with such a father and garden should develop botanical interests, but in Linnaeus' case the passion became the ruling one of his life. His curiosity about and sensitivity for the natural world was deep, abiding, and genuine. So powerful were these feelings that they often compensated for his professional and personal failings, and when he expressed and explained them even detractors were charmed.

Christine Linnaeus was reasonably supportive of the botanical enthusiasms of her husband and son, but she also insisted that Carl follow in the family tradition and become a clergyman. Accordingly, at seventeen he was sent to a local secondary school to commence his preministerial studies. There he did well in mathematics and physics, brilliantly in botany, and abysmally in all subjects related to theology. Finally, one of his principal professors, Johan Rothman, advised the father that his son should be permitted to drop his theological course. However, Rothman said, the boy appeared to be something of a genius at natural history and every effort should be made to allow him to pursue these studies. Rothman himself volunteered to tutor Carl privately and allow him to attend his own public lectures without paying fees.

Nils and Christine accepted Johan Rothman's offer, and eventually Linnaeus left home to enroll as a university medical student. Probably this was something of a disappointment to the parents, for in those days, while clergymen might be impecunious, they were regarded as respectable professional men, thought leaders, and community pillars. Physicians, on the other hand, were in a class of barbers, midwives, and astrologers. However, for one of young Linnaeus' bent, there was the advantage that, to the extent natural science instruction was thought necessary, it was offered by the faculties of medical schools. This was particularly true of botany, physicians then being much more herbalists than they are today. So in 1727, when he was twenty years old, Linnaeus enrolled as a medical student at the University of Lund.

When he left home Linnaeus was a good, all-around naturalist, perhaps already the most knowledgeable botanist in Sweden. Also, he was very poor, and this seems to have marked Linnaeus more than most. As a young man he was forever scratching and hustling to survive, and even when older, well established, and affluent he never lost his fear of poverty. When he was the most noted naturalist in Europe and a consultant to the Royal Zoological Gardens in Stockholm, a Dutch collector sent Linnaeus a list of wild animals he had for sale. Linnaeus admitted there were some very attractive creatures on the list, anteaters and such, but he remarked, "My hair stands on end and lice bite at its roots when I look at the prices in the catalog: 300, 100, 50 gulden—what one has to pay for a pair of riding horses without a coachman. All animals are beautiful, but money is even more beautiful."

Throughout his life Carl Linnaeus was to be a true lover of the beauties of the Goddess Flora, but he also showed true respect for the power of Mammon.

To stay at Lund, Linnaeus needed what would now be called a grant. In those days this involved finding a patron. Linnaeus soon did, attaching himself to Kilian Stobaeus, a university lecturer, local physician, natural history collector, and bibliophile. They got off to a bad start when Stobaeus caught Linnaeus late one night using, uninvited, his private library. However, after his first anger passed, the physician decided the young man had a genuine interest in the books. He not only forgave the intrusion but gave him free room and board. "He loved me not as a pupil but as a son," Linnaeus later remarked, but did not add that this affection lasted for a very brief time. After a few months Linnaeus concluded he could do better than either Lund or

Kilian Stobaeus and left both in the spring of 1728. Stobaeus thereafter considered him an insufferable ingrate.

Throughout his career Linnaeus was adept at getting and playing patrons, but his relationships were consistently like those with Stobaeus. Linnaeus would use their resources and influence to suit his needs—until he got a better offer from a greater man. Some of the discarded patrons remained professional admirers, but few retained much personal affection. Though he lived more than seventy years, Linnaeus seems to have had very few old friends.

In the fall of 1728 Linnaeus enrolled in the medical school at Uppsala, the most prestigious university in Sweden. There he improved himself considerably as far as patrons were concerned, catching the attention of two important ones. The first was Dr. Olaf Celsius, professor of theology, dean of the cathedral, and an impassioned botanist who specialized in the study of plants mentioned in the Bible. A nephew of Celsius is remembered as the inventor of the thermometer of the same name, though Linnaeus may, in fact, have designed the first such instrument.

The second was Olof Rudbeck, then perhaps the best-known scientist on the Uppsala faculty and the son of an even more famous father. The elder Rudbeck had been a pioneer in the systematic study of human physiology and also had compiled a massive catalog of what he believed were all the known plants of the world. This work apparently was similar in scope to that which Linnaeus was to undertake later, but the manuscript was destroyed in a fire prior to its publication and the elderly author had neither the spirit nor the time to re-create it.

Celsius and Rudbeck took turns feeding and housing Linnaeus and promoted his career. Between them they were able to obtain a small royal scholarship for their protégé and then, though he was only a second-year student, arranged to have him appointed as a botanical lecturer in the medical school.

One reason for their high opinion was that in 1729 Linnaeus had written a provocative paper, which—with a keen instinct for what side his bread should be buttered on—he had dedicated to Dean Celsius. It was titled *Praeludia Sponsaliorum Plantarum* and had to do with the sexuality of plants. A few previous academics had dealt with the subject—though a bit gingerly—and in nonscholastic circles it had, of course, been recognized for centuries. In the body of his sex paper, Linnaeus made some sound and newish observations about stamens, pistils, and other botanical phenomena, but he commenced in an exceptionally sensuous way:

"In spring, when the bright sun comes nearer to our zenith, he awakens in all bodies the life that has lain stifled during the chill winter. . . . Words cannot express the joy that the sun brings to all living things. Now the black-cock and the capercailzie begin to frolic, the fish to sport. Every animal feels the sexual urge. Yes, Love comes even to the plants. Males and females, even the hermaphrodites, hold their nuptials (which is the subject that I now propose to discuss), showing by their sexual organs which are males, which females, which hermaphrodites. . . .

"The actual petals of a flower contribute nothing to generation, serving only as the bridal bed which the great Creator has so gloriously prepared, adorned with such precious bedcurtains, and perfumed with so many sweet scents in order that the bridegroom and bride may therein celebrate their nuptials with the greater solemnity. When the bed has thus been made ready, then is the time for the bridegroom to embrace his beloved bride and surrender himself to her."

Later Linnaeus was to expand this student paper and incorporate it into his general system of classification. But its immediate effect was to make him quickly prominent, not entirely for scientific reasons, within the small gossipy European natural history establishment. The reaction—including that of Dean Celsius—was that while the young man's prose might be a bit purple his observations and ideas were interesting and important. However, on the fringes there were those who responded like a hard-shelled Baptist preacher might to the announcement that nudists were planning to camp in his part of Arkansas. Among these was Dr. Johann Siegesbeck, director of the botanical gardens the Russian czars had established in St. Petersburg. Among other things Siegesbeck said that such "loathsome harlotry as several males to one female would not be permitted in the vegetable kingdom by the Creator." He also raised—angrily—a good many rhetorical questions: "Who would have thought that bluebells, lilies, and onions could be up to such immorality? How could so licentious a method be taught to the young without offense?"

Siegesbeck was able to protect literate Russians from Linnaeus by having his paper banned as "lewd and licentious," but in general his criticism was regarded as a display of botanical know-nothingism. In time Linnaeus had his revenge. He named a small-flowered, "unpleasant," and burry weed *Siegesbeckia*. Later still, as taxonomical recreation, he published a paper in which he classified practicing natural historians in a military hierarchy according to their merit. At the top of the heap was, of course, "General" Linnaeus, followed by a number of naturalist colonels, majors, and captains. The last man listed was "Sergeant Siegesbeck."

In those days, would-be professional naturalists were expected not only to publish but also to make at least one exploratory collecting expedition. Linnaeus decided upon Lapland in northern Scandinavia. Others, including the younger Rudbeck, had visited the region briefly, but it was assumed that much remained to be discovered there.

Beyond scientific interest there were apparently compelling social reasons for Linnaeus' leaving Uppsala. He had been asked to give up his lodging with the Rudbeck family, as he had become involved in a messy triangle with his professor's wife and a female friend of hers. Rumors abounded, but the only surviving account of the situation is a very brief comment by Linnaeus, who wrote: "The unfaithful wife of Librarian Norrelius now visited the Rudbecks, with the result that Linnaeus became the object of such hatred to his hostess that he had to leave."

Whatever the difficulty, it did not sour the professional relationship between Linnaeus and Olof Rudbeck, who helped arrange for a small grant from the Swedish Royal Society of Science to help pay for the Lapland trip. Later, Linnaeus responded by naming a plant after his mentor, in this case a very nice one, the black-eyed Susan. A fairly accurate catalog of Linnaeus' friends and enemies can be made by referring to what sorts of plants he named for whom.

In the late spring of 1732 Linnaeus started north, riding a cheap horse and carrying, among other things, a leather bag, two extra shirts, a shag coat, short sword, fowling piece, plant press, notebooks, two nightcaps, a green fustian hat, and a pigtailed wig. He stayed in the field for four months and discovered a good many things new to European science. He filled his notebooks with not only biological but anthropological observations, being alternately fascinated and repelled by the Lapps.

Although he had a few days of good weather and easy traveling, which were a pure joy when "the nightingale sang to him all the way," Linnaeus generally was wretched, continually complaining in his journal about bad weather, bugs, food, and travel conditions. If nothing else, the experience in Lapland seems to have convinced Linnaeus that fieldwork in uncivilized places was not his cup of tea. Later he had many invitations to make spectacular trips to improbable parts of the globe, but he always turned them down and confined his field trips to the settled regions of southern Sweden.

Toward the end of his summer in Lapland there occurred an incident which illustrates how he felt about this trip and also gives some insight into his character. In July, at Sorfold, a fishing town on the Norwegian coast about a hundred miles south of the modern city of Narvik, Linnaeus heard about an interesting fell region in northern Sweden. He apparently decided that visiting this place would make his Lapland tour sound much more impressive. However, the area, Torne Fells, lay some three hundred miles inland, and by then Linnaeus had had enough of roughing it. So he wrote and later published an entirely fictional account of a trip to the region. He invented details about a mighty trek of some eight hundred miles, which he said he completed in fourteen days. This story was accepted as fact, since nobody knew much about Lapland, but later biographers nailed him in this lie, and it has raised justifiable doubts about his integrity.

Linnaeus returned to Uppsala in October 1732 and pried a bit more money out of the Royal Society on the grounds that his trip—including the excursion to Torne Fells—was much more extensive than originally planned. However, in his absence, rivals had undercut his reputation and appropriated his academic stipends. For the next two years Linnaeus was forced to tutor private pupils, lead student study parties, and wander about southern Sweden giving lectures in botany and geology. He spent the winter of 1734–35 in the mining town of

Falun as a guest of the provincial governor. There he met Johan Browallius, then the chaplain of the governor's household and later bishop of Abo. Linnaeus unburdened himself to Browallius, saying he was terribly depressed by his lack of professional recognition and his chronic poverty. The young clergyman advised him to marry a rich woman "who could first make him happy and then he her." Linnaeus wrote that the suggestion was one which he would never seriously consider acting upon. But two months later he became engaged to Sara Elizabeth Moraea, daughter of one of the wealthiest men in the provinces.

Sara Elizabeth was called the Fair Flower of Falun, but in surviving portraits she looks to be a rawboned, grim-mouthed, chinless young woman. Later, acquaintances of the botanist were virtually unanimous in the opinion that she was a small, mean-minded woman and a notable termagant. Her father, Johan Moraea, may have recognized that she was not going to be the easiest daughter to settle and could not have had many illusions about why Linnaeus was willing to take her off his hands. However, he decreed that he would give the hand and dowry of Sara Elizabeth only if Linnaeus obtained a medical degree with which he could earn a living. For his training he went to Holland, where there were medical diploma mills that attracted would-be physicians from all over Europe.

Linnaeus stopped first in Hamburg, where he received some favorable publicity as the intrepid explorer of Lapland. However, he was forced to leave the city hurriedly because he gave the opinion that a famous stuffed hydra owned by the burgomaster, who was trying to peddle it at a large price to the King of Denmark, was a fake, manufactured by monks out of yards of weasel and snake parts. In June he turned up in the Dutch university town of Harderwijk, bearing with him a previously written medical paper in which he developed the theory that people who lived in regions of heavy clay soil were particularly prone to fevers. Within one week the dissertation was published and defended before a faculty committee and Linnaeus was granted a Doctor of Medicine degree. This chore painlessly completed, he was free to pursue larger interests.

This was the golden age of Holland, a time when Dutch sailors and merchants were ranging the world, accumulating large fortunes for themselves and their backers. Many were intensely interested in natural history. A journalist of the time commented, "Practically no captain, whether of a merchant ship or man-of-war, left our harbors without special instruction to collect everywhere seeds, roots, cuttings, and shrubs and bring them back to Holland."

The newly rich established elaborate private zoos, gardens, and museums to display the curiosities, and there was a profitable trade in them. There also was a great demand for natural history technicians who could maintain, catalog, and publicize these collections. Linnaeus immediately began wheeling and dealing in patrons, and within two months after receiving his medical degree he found George Clifford.

Clifford was an English banker who, as an investor in and director of the Dutch East India Company, had become one of the richest merchants in Europe. He had settled in Holland and built a great estate at Hartekamp that included gardens, greenhouses, a zoo, and a library, which made it one of the great natural centers of the day. After first seeing the place, a dazzled Linnaeus wrote to Clifford:

"When, finally, I entered your truly regal house and splendidly equipped museum, whose collections speak no less of their owner's renown, I, as a foreigner, felt quite carried away, for I had never seen their equal. I desired above all things that you might let the world have knowledge of so great an herbarium, and did not hesitate to offer to lend a helping hand."

Clifford was willing to be a patron and, as a roaring hypochondriac, found the prospect of having both a physician and naturalist on his staff very attractive. However, there was one difficulty. Prior to visiting Hartekamp, Linnaeus had made a firm commitment to Johannes Burmann, director of the Amsterdam Botanical Gardens, to work with him on a floral catalog. Clifford solved this problem quickly. Burmann had long coveted a rare and expensive book, *The Natural History of Jamaica,* of which Clifford had two copies. He wrote Burmann, "I will give you one of them if you will give me Linnaeus."

The trade was made, and in September 1735 Linnaeus moved to Hartekamp, where he was to remain for most of the next two years. He was paid a thousand florins a year, probably more than he had earned in all of his previous working years, and given an expense account for research and travel. He was established in a sumptuous private apartment, coddled by servants and retainers. While with Clifford, Linnaeus turned out a prodigious amount of work, which both publicized his patron's collection and established his own reputation.

During his Hartekamp period Linnaeus published a dozen substantive papers and books. Three of them, *Systema Naturae, Fundamenta Botanica,* and *Genera Plantarum,* set forth a solution to what he perceived to be the critical problem of contemporary natural science. During the rest of his life he was to expand and revise the Linnaean method of classification which these papers outlined, but with these publications the system was on record and can be considered as a whole.

The problem was that while a considerable body of natural history data had been accumulated during the previous two millenniums, it existed in a great jumble. Neither scholars nor laymen had a clear idea of what, if any, relationships there were between the parts of the disorderly whole. Linnaeus began sorting through and straightening up. He concluded, rightly, that until this heavy taxonomic work was done any systematic investigation of the nature of nature was impossible. He described his chores by saying, "God creates, Linnaeus arranges." the blatant hubris of the statement may still make teeth grate, but it is a fair summary of his purpose and the mind-boggling extent of his life's work.

A major cause of the confusion was the lack of a generally accepted nomenclature for the natural sciences. "If you know not the name, knowledge of things is wasted" was a classical truism of which Linnaeus was well aware. And if nomenclature is so difficult or obscure that only a few can master it, a society will invariably be authoritarian and elitist, a phenomenon that has been noted by many social philosophers.

Obviously, in Linnaeus' time everything that was known had a name. Most plants in a given area had serviceable common names, but these often varied within even a small region, to say nothing of when the ranges extended across linguistic boundaries. Some plants had by then been examined and identified with Latin or Greek labels. However, these were customarily lengthy descriptions, listing numerous characteristics rather than working names. For example, scholars knew a plant by the name *physalis amno ramosissime ramis angulosis glabris foliis dentoserratis*. This, according to Joseph Katner in his history of natural history, *A Species of Eternity,* can be translated as a "bladder-fruited annual, many-branched with angled branches and smooth, deeply toothed leaves." We know the plant as ground cherry.

Surveying the chaos, Linnaeus suggested that natural creations be given names that would consist of two words. These should be classical, preferably Latin, since this was the nearest thing to a universal written language in Europe. The first word should indicate genus and place it in a group of similar things. The second should describe an obvious characteristic that set it apart from other members of the genus—made it a unique species. Under the Linnaean system the ground cherry formally became *Physalis angulata,* that is, a member of a genus of plants that bear bladder-like fruit but one that has sharply angled leaves.

The key elements of the system were not original. Common names were, and still are, largely binomial: White oak, skunk cabbage, ground cherry. Combining the succinct common nomenclature with the descriptive, ordered classical one to create useful, memorable code names may not seem like a mighty intellectual feat. However, others had been struggling with the problem since the time of Aristotle without much success. Furthermore, nobody since Linnaeus has been able to do better.

Binomial classification gave Linnaeus and those who followed a means of identifying and cataloging individual species and generic groups. However, there remained the larger disorder, comparable to a bureau in which socks, shirts, and underwear are stored in random piles. Compartments were needed, in which everything could be arranged in logical relationship to everything else. Linnaeus began by designating three natural kingdoms: minerals (which grew); plants (which grew and lived); and animals (which grew, lived, and felt). These would be subdivided into classes and orders—and then genera and species.

Linnaeus first applied the system to the plant kingdom and, in looking for a means of separating classes and orders, returned to his early interest in floral sex. Class would be determined by counting the number of stamens in a bloom, and orders by the number of pistils. Both class and order names should reflect these numbers: Monandria (one stamen), Diandria (two stamens); and Monogynia (one pistil), Digynia (two pistils), etc. Placing a plant in the proper category became a simple arithmetical matter. A canna, for example, with one stamen and one pistil, was of the class Monandria and the order Monogynia.

In theory, distinctions between genera were also sexually determined, based on characteristics of the fruit. However, in practice, Linnaeus and the Linnaeans gave themselves considerable latitude, often using vegetative characteristics, classical allusions, and, of course, the names of both friends and enemies. Names had to be learned by rote, and Linnaeus stated that a master botanist should know most of the classified species—which at the time numbered about 6,000.

The new system was much more comprehensive and convenient than any that had existed before. After its introduction, John Bartram, a Pennsylvania farmer; Peter Collison, a London merchant-horticulturist; and Professor Carl Linnaeus could exchange information about plants whose names and general characteristics were familiar to all. Almost any observer who had a little Latin could pick up a plant, see that it had five stamens and one pistil, and therefore belonged in the class Pentandria and order Monogynia. If it had a bladder fruit and angled leaf it was either *Physalis angulata,* the ground cherry, or something closely related to it.

Stamens and pistils, however, are secondary floral characteristics, and their number does not necessarily signify close relationships between plants. Simply counting them as a means of defining classes and orders led to some ludicrous arrangements: for example, putting basswoods, poppies, and water lilies in the same class and the mints and sages in separate ones. So even before the master died, others were rearranging the Linnaean classification using vegetative characteristics to define categories. By the early nineteenth century the rigid, arithmetical system based on stamen and pistil counts had been discarded.

In fairness, it should be noted that Linnaeus himself frequently noted that his system was arbitrary and artificial. However, he contended, the immediate need was *some* system, not necessarily a perfect one. Later, he suggested, more precise classifications could be made. It would be a long time, he added, before anyone knew enough about nature to devise an entirely natural taxonomy.

In this, Linnaeus was prophetic. Counting stamens and pistils may be obsolete, but our species, genera, orders, and classes are still abstractions. They do not exist in nature, but serve as code devices which enable us to think more easily about nature.

One of the more extraordinary things about the Linnaean classification was how rapidly it was adopted. Linnaeus published *Systema Naturae* in 1735 and the other critical papers during the next two years. Within a decade, his classification had become the standard. Few scientific or intellectual innovations have been so quickly accepted. The advantages and convenience of his method, even if imperfect, were apparent. Also, though there has been reluctance to classify Linnaeus as a genius, he was at least a true maestro when it came to nomenclature, and he named some 12,000 plants and animals. He was a quick, clever observer with a good eye for significant detail and a flair for free-form association. The names he gave things not only were sensible and memorable but also had a certain panache.

Finally, the speed and completeness with which his system was accepted reflected certain elements in Linnaeus' character. The Linnaean system might not have been so quickly and generally recognized had the author been a less arrogant and aggressive man. A diffident Newton or Darwin probably could not have done so rapidly what Linnaeus did. In effect, he knocked together the heads of natural scientists and said: "As of now we are going to do things my way."

In 1738 Linnaeus, who was increasingly being badgered by his fiancée and her father, returned to Sweden. Though he was already a great figure in the natural history field elsewhere in Europe, his fame had not spread to Sweden, and he found it necessary to commence practicing medicine in Stockholm. He began by treating young city gallants for venereal diseases, using a cure he had obtained from a French physician. Linnaeus had great success, and was shortly treating forty to sixty patients a day.

Within a few months he improved not only his fortune but his class of patients. He began treating the gentry and even royalty for more mentionable diseases. In the spring of 1739 he finally married Sara

Elizabeth. Thereafter he was never poor again. Also, in 1741, after some brisk academic skirmishing, he returned triumphantly to Uppsala University as a full professor. This was to remain his working base for the rest of his days. In a rare burst of genuine, first-person enthusiasm he commented on this development:

"By God's grace I am now released from the wretched drudgery of a medical practitioner in Stockholm. I have obtained the position I have coveted for so long: The king has appointed me professor of medicine and botany at Uppsala University. . . . If life and health are granted to me, you will, I hope, see me accomplish something in botany."

Linnaeus continued as the major academic star of Uppsala for the next thirty-seven years—showered with professional and social honors and attention, which bathed but did not noticeably diminish his ego. Eventually he was knighted by the Swedish king, and thereupon took the name Carl von Linné. Also, his work was generously supported by private and public patrons. Prominent friends directed him toward profitable real estate investments. Two ventures in applied natural science—developing a technique for artificially stimulating North Sea oysters to produce pearls and the establishment of a Swedish tea plantation—were successful financially, though not artistically.

In regard to his health Linnaeus was less fortunate. He remained capable of prodigious bursts of energy and accomplishments, but these were increasingly interspersed with periods of illness and depression. At times even the smallest tasks seemed impossible. He suffered from severe migraine headaches, rheumatism, mysterious fevers, and in his last years from a series of strokes. He referred to the first of these, which struck in 1774, as a "message of death." The second occurred two years later and left him almost totally incapacitated. He died in the early winter of 1778.

Many students and biographers of Linnaeus consider him a hypochondriac, or even a manic-depressive or schizophrenic; perhaps he suffered from psychosomatic ailments. Such exercises in historical diagnosis are speculative at best. However, Linnaeus' own writings indicate that he experienced substantial inner tensions that may well have affected both his physical and mental health.

Perhaps this tension reflected Linnaeus' growing realization that his work in natural history had put him at odds with the rigid theological world-view that had dominated his youth and the European intellectual life for centuries. For example, Linaeus concluded in private writings and conversation that many biblical premises having to do with creation—the length of time it took and the fixed nature of worldly phenomena—were inaccurate. He never espoused such iconoclastic theories in public, but they were implicit in much of his work and many of his lectures. Daring and rebellious students picked up on them and began following them to logical if, by the standards of the time, heretical conclusions. On several occasions hard-line theologians suggested that Linnaeus was laying himself open to charges that he was a closet atheist. These warnings alarmed Linnaeus, perhaps because he was a man who valued conventional acclaim, respectability, and affluence. On at least one occasion he took aside a controversial student and begged him "never to dispute with theologians again in this fashion, since they would never change but would grow to hate natural history, and that would be very harmful to the latter."

Linnaeus practiced what he preached. Not only did he avoid stirring up controversy, but he publicly took conservative intellectual and religious positions. During the latter half of his life he worked sporadically on a strange manuscript, "Nemesis Divina," which he never published. The central thesis was that God, somewhat like an indefatigable, irascible private detective, was forever ferreting out the transgressions of mortals and continually updating dossiers on them. When sufficient evidence of wrongdoing was accumulated, immediate and often cruel punishment followed. "Nemesis Divina" was a lengthy listing of examples of God's vengeance, drawn from Linnaeus' own experience, folk tales, and classical histories. It was comparable to the black, hopeless essays and fictions which Mark Twain composed toward the end of his life and was, as in Twain's case, aberrant in regard to the body of Linnaeus' work. However, the grim manuscript suggests the torments Linnaeus experienced because he seems to have known he was a principal figure in the transition between old and new modes of thought and belief.

Happily, in regard to his final wish that he "accomplish something in botany"—and, by extension, natural history—his heart's desire was fulfilled as it has been for few other men.

After his appointment to Uppsala and until his death Linnaeus was occupied largely with revising, expanding, and applying the systems of nomenclature and classification he had developed earlier. It was an important chore, and he did it well. But it was repetitious rather than creative work, and his scientific scope and skills did not expand greatly. However, his reputation did—dramatically. Few men have so dominated a substantial discipline as Linnaeus did natural sciences in the mid-eighteenth century. As God's Registrar, one of his most splendid sobriquets, he certified not only natural wonders but the reputations of naturalists. They vied with each other to bring specimens and observations to the attention of the great man who had invented the marvelous new taxonomy. If they were accepted and acknowledged, their own status automatically improved. Students flocked to Uppsala and Linnaeus. The best and brightest of these students sallied forth from Uppsala on bold collecting quests to all parts of the world. As they traveled they converted others to the Linnaean systems, spread the fame of their master, and sent specimens and notes back to him.

Though Linnaeus was no populist, he was the greatest popularizer of natural sciences that the world has ever known. This accomplishment may have been more important to Western culture than his scientific ones. For the next one hundred and fifty years natural history was to engross members of the more affluent and intellectual classes. Many were dilettantes caught up by a popular fad, but Linnaeus also turned on some of the best minds of his time to the joys and mysteries of nature, gave them techniques for systematically accumulating knowledge, and turned them loose on the world.

Though botany was unquestionably his forte, Linnaeus was also a competent and observant zoologist. In fact, his classification of animals seems sounder than his botanical one—or at least more natural. For example, in the class Mammalia his first order was that of the primates. In it by reason of their apparent relationship he places apes, monkeys, lemurs, and man.

With the advantage of great hindsight it seems as if some of Linnaeus' observations should have led him to challenge the prevailing view that nature had been created all of a piece and was forever fixed. That he did not—did not in effect become Charles Darwin—is at the root of much of the later criticism of his science and philosophy. However, to a greater degree than is often acknowledged, Linnaeus' systematic cataloging of the objects of nature laid the essential groundwork for the revolutionary discoveries and theories of the next two centuries. Perhaps even more important—as the unrivaled popularizer of the natural sciences—he made them inevitable.

Great Bears and People

If individual human personalities could be stuffed and displayed like the carcasses of trophy gorillas or reconstructed like the skeletons of brontosauruses, Henry Kelsey would make a fine display for a major museum. He was a courageous soldier, an effective diplomat and a successful entrepreneur. In terms of his own time, the 17th century, he was a keenly observant, progressive-minded naturalist, cultural anthropologist and topographer. In terms of all time, Kelsey was an explorer of the very first rank. Before reaching adulthood, he saw more of this continent when—for Europeans—it was absolutely virgin than any other white man had or would. He was also a poet, not a very good one but perhaps the first English-speaking individual with the temperament and talent to be a bard of the true, howling wilderness of the New World.

Kelsey arrived in the New World in 1685 as a 14-year-old boy-of-all-work who took part in the commercial buccaneering expedition that drove the French out of northern Ontario. He remained as a "servant" of the newly chartered Hudson's Bay Company. During the next five years he became, by the standards of the firm, odd; the only one of that hard crew sufficiently interested in the "red niggers" of the northern wastelands to learn anything about their language and lore or, it seems, to conclude that they were indisputably human.

In 1690 Kelsey was sent off from York Factory—a fortified post at the confluence of the Nelson and Hayes Rivers—on an extraordinary mission that was to keep him in *terrae incognitae* for the next three years. Officially he was ordered to establish a fur trade with savages who, it was assumed, might be found in the western interior. The masters of the Bay Company may also have concocted the assignment as a means of ridding themselves, at least temporarily, of a restless young man of unsettling, uncivilized opinions. In any event, Kelsey was very likely the only man who would have made such a trip. He left with a party of Cree tribesmen and with a "fire in his heart" to see the lands beyond the edge of the known world. When the Crees dared go no farther, he joined other Indian hunters, one of whom thought Kelsey insane because he "was not sensable of danger."

Always heading west, Kelsey skirted the great Arctic Barrens, then slanted southward across prairies, traveling at least as far as the eastern foothills of the Shining Mountains—the Rockies. He was the first Englishman to come upon a musk ox and the Great Plains, the homelands of the Sioux, Cheyenne and Crow Indians. Probably in the spring of 1691 he met what he accurately reckoned was the most formidable beast in the New World. It was, he noted, "a great sort of a Bear which is bigger than any white Bear [polar bears were common around Hudson's Bay] & is neither White nor Black But silver hair'd like our English Rabbit."

So impressed was Kelsey, the first of us to see and describe a grizzly, that he summoned his Muse and wrote:

> . . . an outgrown Bear wch is good meat
> His skin to gett I had used all ye ways I can
> He is man's food & he makes food of man. . . .

Subsequently, the grizzly has wasted enough trappers, lumberjacks, drovers and tourists so that the substance, if not the exact words, of the poet's bottom line hasn't been forgotten—that sometimes we may eat this bear and sometimes the bear eats us. Much lesser creatures of the bee, spider or snake sort have proved more deadly, to judge from the available statistics, but only the great bear, when so inclined for defensive or predatory reasons, has been able to make meat of us in direct, *mano a mano* confrontations.

As we began to settle in big-bear ranges, the animal proved even more ready and able to make meals of our cattle, sheep, horses, beehives, orchards and storage caches. It was quickly decided that clearing out the grizzlies was an imperative chore if the land from the plains to the Pacific was going to be a nice place for us to live. This has been largely accomplished. It's thought that there were once 100,000 grizzlies roaming throughout the trans-Mississippi West. Now it's more precisely estimated (counting grizzlies has become something of a cottage industry) that only from 800 to 900 of them survive in the lower 48, mostly in northern Montana, northwestern Wyoming and the Idaho panhandle.

Only raving nostalgics can fail to understand why this has happened or argue seriously that it's possible or desirable to return—bearwise—to the good old days. For example, another guess has it that northern California supported 10,000 of these beasts in the early 19th century. Now there are none, basically because you can't have that many—or, in truth, any—grizzlies and, say, Sacramento. It goes far beyond direct conflict—bears chasing state senators around Capitol Plaza or harassing midnight dope dealers in the bushes along the American River. All that bears imply in terms of habitat and ecology, and all that Sacramento implies in the way of human use and logistics make the species and city absolutely incompatible.

In their behavior toward us, grizzlies haven't changed appreciably since Kelsey met his first one. They remain determined to pursue and defend their own self-interest, by violent means if necessary, and they generally don't give a damn about what we want. There's no reason to expect they can be persuaded or forced to act differently than they always have.

On the other hand, our opinions about grizzlies have changed. There are still a few holdouts in bear country—for obvious reasons—who cling to the old view that the only good grizzly is a dead one. However, the prevailing attitude now clearly is that we should cherish our remaining grizzlies because this is in keeping with contemporary environmental sensitivities and because they are rare, esthetically impressive creatures, appealingly associated with our national history, or at least romantic versions of it. There is also some chauvinism involved. Passionate pro-bear people sometimes give the impression that the few animals remaining in our northern Rockies are the last of the species on the continent. This isn't the case. There are between 40,000 and 50,000 grizzlies in Alaska and western Canada, and the evolutionary future of the bears lies there.

What it boils down to is that around 1970 we collectively decided that, for essentially sentimental reasons (which shouldn't be equated with foolish or trivial ones), the grizzlies remaining in the lower states were desirable adornments and should be preserved. Political authorities sensitive to such issues gave the appropriate public servants, wildlife and land managers, the job of executing the will of the citizenry, and ever since, we've been squabbling about how to do this work and why it hasn't been done faster and better. If nothing else, we've come to appreciate that there are few things in the world so difficult to preserve as a wild grizzly bear.

For the most basic physical reasons, a grizzly is a great deal more difficult to manage and manipulate than a black-footed ferret or a California condor. The species isn't senescent, clinging to a small scrap of prehistoric habitat, unable to survive independently in and cowed by the world as it now is. Rather, the remaining grizzlies are robust, adaptive omnivores who, if permitted, are quite capable of taking care of their own bed and board almost anywhere in their former range, i.e., the western two-fifths of the country. They don't even require excessively wild habitat, having frequently demonstrated that they're able and willing to tolerate our presence and, in fact, are fairly ingenious about finding ways to benefit from it. Their principal trouble is that we still find it difficult to adjust to free-ranging grizzlies. When it gets down to the nitty-gritty of how and where to keep such creatures, our good intentions run smack into the Sacramento syndrome.

To show concern for another species, up-to-date wildlife scientists almost reflexively try to put collars, bearing radio transmitters, on the creatures in order to follow and collect information about them. This is called research, and it sometimes is beneficial to the animals but, if not, at least it gives a soothing sense that good works are being done on their behalf. The grizzlies have come in for a lot of this sort of attention. Some 30 to 40 of them (from 4% to 6% of all the wild ones in the West) now wear transmitters. One of the wired bears is No. 38, a mature female which lives in the Yellowstone ecosystem, a tract of more than 8,000 square miles, about 40% of it within Yellowstone National Park; the remainder is composed principally of adjacent national forest lands.

Late last summer, Grizzly 38, accompanied by two cubs, began traveling in a southwesterly direction away from national park land toward the Tygee Basin area of the Targhee National Forest in Idaho. On Aug. 23 her transmitter signal was picked up by Dr. Dick Knight, a veteran Yellowstone bear biologist who was flying over the area, as he frequently does, to monitor bears equipped with transmitters. Knight called George Matejko, a Forest Service ranger–wildlife manager who works the Targhee Island Park District, and told him the three bears were headed his way. Matejko then called Bill Enget, a rancher who holds a federal permit to graze sheep on 28,126 acres of forest land in Idaho and Montana. That day Enget had about 2,000 sheep in the Tygee Basin being watched by two shepherds, Jess McIntire and Jose Rodriguez. As Enget had previously been informed, the sheep were then in one of the areas that, to protect grizzlies, federal agencies have designated as Situation I lands. In Situation I areas the welfare of grizzlies comes first, those of human users and property owners second. (There are also Situation II and III areas in which, vis-à-vis bears, human interests are progressively upgraded.)

Enget said that though his permit entitled him to several more weeks of grazing on the land, he would immediately start moving the sheep back to a shipping corral at Big Springs, some 15 miles distant, to avoid any trouble with the bears. However, before anything could be done, No. 38 and her cubs arrived and jumped the sheep on their bed-ground. They were driven off by warning rifle fire from McIntire and Rodriguez, but during the night two lambs were injured and later had to be destroyed. (Before the episode was concluded—18 days later as it turned out—Enget lost 33 sheep. A Forest Service report says only four of them, beyond doubt, were killed by the bears; 13 others were reckoned to have been done in by coyotes or poisons; the remaining 16 died of causes about which the concerned parties could not or would not agree.)

On Aug. 24, Enget, his wife, Louise, and two nephews, Hal and Jeff Buster, joined the shepherds and started driving the flock. Shortly, a Forest Service biologist, Barbara Franklin, arrived on the scene and began riding around the herd looking for bear sign. None was found, and for the next three days nobody knew where the bears were, the weather having turned so bad that Knight couldn't fly. On the 27th, visibility improved, and Knight got a signal which indicated that No.

38 was only a half mile from the flock. That night, despite patrols by the herders, another lamb was killed.

Two days later, while working 40 sheep that had split from the main flock, Hal Buster, on horseback, was charged by No. 38, probably in defense of her cubs. Buster fired shots over the sow's head, and she retreated without doing any damage. During the next few days, the Forest Service closed several nearby campgrounds and evacuated two of its own nonbear work parties and some commercial loggers and sawmill operators. As these bystanders left, a mixed company of 20 bear people—U.S. Fish and Wildlife, Forest Service and Idaho state game agents—assembled. Then some members of the press (who, having learned of the happening, followed it enthusiastically) joined the ground party, which was in considerable disagreement about what should be done. Enget, backed by other ranchers after they heard the details, wanted the bears removed—lured or trapped—before more sheep were lost or people hurt. Fish and Wildlife agents argued that this was Situation I country, where it was people and sheep who had to be moved, not bears. Forest Service operatives were indecisive, torn between their obligations to the grizzlies and to Enget. Idaho game officers, who had bear-catching and -handling equipment, said they would use it, but not until the Feds made up their minds about what to do.

A compromise, which didn't entirely satisfy anyone, was struck. Enget would continue to move his sheep out of the area, and the agents would try to cover his retreat. Attempts were made to haze the bears with aircraft. Propane cannons (noisemakers often used to frighten birds out of orchards) were brought in and fired at the grizzlies, as were flares. On several nights Idaho game agents strung as much as a half mile of electric fence around the sheep bed-grounds. The three bears proved to be singularly unshockable. They had no difficulty keeping up with the slow-moving sheep, which could be driven only two miles or so a day, hanging around the flock and picking off a few more lambs.

The improbable entourage of herders, public servants, press, sheep and bears arrived at Enget's Big Springs corrals on Sept. 4. There the rancher sold his lambs, several weeks earlier than he had planned. He then drove the ewes and bucks back to his ranch, two miles away. The bears followed. All parties then milled about for several days, and the grizzlies began touring private and federal lands on which their status was less protected than in Situation I areas. It was decided that to avoid a serious incident and/or vigilante action by aroused local residents, the family had to be removed. Three culvert traps were set and baited with bacon, fruit and a dead sheep supplied by Enget. This stew was doused with vinegar, which smells like formic acid produced by ants—on which the grizzlies had been observed to snack between sheep meals. The bears roamed about for another day or so and at 1:30 a.m. on Sept. 9 entered the traps and were taken. Subsequently they were tranquilized, trucked back to and released in Yellowstone Park. Except that the sow got a replacement radio collar and the cubs their first ones, the three bears were unmarked by their excursion. On the human side, the incident cost about $15,000 for sheep, salaries, aircraft use, electric fencing, propane cannons and the preparation of a report.

Grizzly No. 15, a 430-pound male, was first radio-collared in 1975. During the next nine years he got into no trouble with people but showed a marked inclination to forage in the Gallatin National Forest on the northern (Montana) edge of the Yellowstone ecosystem. (Grizzlies may travel 100 miles a day while taking care of business.) He was in the Gallatin on June 24, 1983. Early on the morning of the 25th, he padded up to a Forest Service campground at Rainbow Point, where William May, 23, of Sturgeon Bay, Wis., was camping with a friend, Ted Moore. Without warning, No. 15 charged the tent of the sleeping campers, bit through the side, picked up May, carried the screaming man 30 feet away, killed him and ate about a third of his body.

One of the first outsiders to arrive on the scene was Dr. Chris Servheen, a Fish and Wildlife Service biologist and the chief scientific consultant to a committee of federal and state executives now grappling with the problem of grizzly preservation. As such, Servheen is the ranking grizzly professional among public servants in the country. As to bear No. 15, Servheen had no doubts about what to do, there being a firm, if unwritten, rule that a grizzly that kills a human must be killed immediately or permanently kept as a study captive. Some wild justice—the urge for revenge—underlies this practice, but it is also supported by certain behavioral observations. Grizzlies are very adaptive, i.e., they learn quickly, and if they hit upon a tactic that works for them, they remember and are apt to repeat it.

On the night of June 25, No. 15 returned to the scene of the tragedy and was caught in a foot snare lashed to a stout tree. As the enraged animal was tearing at the trap and its cables, a Montana game agent shot a tranquilizer dart into its side. Thereafter, an analysis of material under the bear's claws and in its excrement proved that this was the right animal, the one that had killed May. Servheen injected a dose of lethal poison into its side. An autopsy was then performed, one theory being that perhaps the animal was suffering from a disease that had caused behavior derangement and, if contagious, might similarly affect other grizzlies in the area. No such pathological evidence was found, nor did the autopsy reveal any abnormalities in No. 15 that might have caused him to act as he did. So far as biologists could determine, this was a robust, sane grizzly.

"A lot of theories have been advanced about why bears turn killer," says Servheen. (Among others: The killers are old or enfeebled animals and thus starving; human sexual or menstrual odors arouse them; uncovered food or garbage triggers a feeding frenzy.) "But none of them applied in this case. The two men had made a good, clean camp in a secure place. They did absolutely nothing—that we can understand, at least—to attract or provoke that animal. This case was a tragic reminder of something we too often forget. We know very little about the inner life, the motivations, of a grizzly. The same can be said about most other animals, but the grizzly is the only one in the terrestrial U.S. about which our ignorance can be fatal. If people and bears are in the same area, I think it is inevitable that there will, occasionally, be tragedies. This has to be a constant factor in our planning about, and the management of, the species."

Alaska aside, our remaining grizzlies principally inhabit two separate tracts, both primarily in large national parks. About 400 to 600 animals live in northern Montana, in and around Glacier Park, while 200 or so survive in the Yellowstone ecosystem. There are probably fewer than 100 elsewhere, in scattered pockets in Idaho, Washington and perhaps southern Colorado. (Until 1980, when an old female was shot in the San Juan Mountains, no grizzlies had been seen in Colorado for nearly 30 years. None has been spotted since 1980.)

In part because they're occasionally joined by bears traveling south from Canada, the northern Montana grizzlies seem to be holding their own. In the opinion of Montana wildlife officials, 25 animals a year can be harvested by sport hunters without destabilizing the population. (Northwestern Montana, outside Glacier Park, is the only area in the lower 48 states where the grizzly is a legal game animal.)

The Montana hunt is intricately supervised. Each year state biologists determine the number of bears known to have died of natural causes or to have been killed in control actions, by poachers or otherwise. This total is subtracted from 25, the number of presumably disposable animals. The remainder, say 15, is what's left for sport. Hunters who draw grizzly permits—about 600 a year—are required to keep game agents informed as to where they are and report their kills immediately. When the designated number of kills is approached, the hunt is closed and field parties are notified. Since it was established a decade go, the system has worked fairly well. Grizzly mortality in northern Montana averages about 20 bears a year—sport hunters taking 10 animals. So long as the overall population doesn't decline, most conservationists accept the Montana hunt as a reasonable tradeoff. In grizzly preservation matters, it tends to maintain the goodwill of hunters and outfitters—a powerful special interest group in this region—and therefore the political benefits for the species as a whole are thought to outweigh the loss of the few animals killed.

In the Yellowstone ecosystem, grizzlies are in a much more precarious position than they are around Glacier. The population has been declining there since at least the mid-1960s, when it was thought that there were from 300 to 400 bears in the area. Now there are from 180 to 210. In the popular sense of the term, the grizzly is considered endangered in Yellowstone. However, under federal law it is designated only as a threatened species. Officially endangered species cannot be molested for any purpose by anyone, including public wildlife managers. The agents who trapped and removed grizzly No. 38 and put down No. 15 after it killed William May would have been guilty of criminal acts if the animals had been legally designated as endangered. The reasoning behind the fact that the grizzly isn't so designated is that if field people don't have some freedom of action, conflicts between bears and humans will increase and worsen, making the species more vulnerable to acts of private vengeance and retaliation.

For the past 15 years, bureaucrats, scientists and private wildlife organizations have been feuding about why the Yellowstone grizzlies are disappearing and who's to blame. The Park Service has frequently been accused of mismanagement, principally because in 1970 and '71 it closed open-pit garbage dumps at which for more than 50 years many grizzlies had fed. Critics claimed that the closures nutritionally impoverished the area and dispersed hungry bears into the fringes of the ecosystem, where they got into more trouble with humans, often with fatal results for the animals. Elsewhere, the Forest Service was charged with being more interested in maintaining cozy relationships with commercial users of its lands—ranchers, timber companies, oil drillers and such—than in accommodating grizzlies and their protectors. State game agencies caught heat for not being energetic enough in enforcing laws that would benefit the bears.

In the 1970s many, not particularly successful, attempts were made to promote peace and cooperation among the disputing authorities. The latest effort of this sort occurred early in 1983 with the establishment of something called the Interagency Grizzly Bear Committee. It comprises upper-level federal executives (regional director types) from the Bureau of Land Management, the Forest, Parks and Wildlife services and comparable representatives from the states of Idaho, Montana, Washington and Wyoming. Servheen is the scientific adviser to the group, in effect its chief of staff.

There has been some understandable skepticism about the IGBC. However, the creators of the new group insist it intends to do substantive work. Among the optimists is Ray Arnett, an Assistant Secretary of the Interior who heads both the Parks and Wildlife services. Arnett says, "Previous committees were technical groups made up of biologists. They did a good job of fact-finding but couldn't make management decisions. They could only report separately to their own agencies. In this new setup, people who can make broad decisions are sitting down together, getting the same technical information at the same time, trying to work out conflict-of-interest problems. I think the willingness to do this is evidence that we've finally all come around to realizing that if we're going to save the animal there's no more time for people to be dancing around trying to protect their own bureaucratic turf. I can guarantee that so far as the Feds are concerned, keeping grizzlies in the Yellowstone ecosystem is now an absolutely top priority.''

Arnett is a big man, bearish himself in appearance if one may say so, who was the chief California wildlife manager when Ronald Reagan was governor of the state. He was appointed to his federal position by former Secretary of the Interior James Watt and in most social and economic matters is a rock-ribbed conservative. However, Arnett, a passionate hunter and naturalist, has demonstrated a visceral compassion for wildlife and wilderness. In specific cases he has proved to be a formidable preservationist, though he would probably rather choke than so describe himself. The grizzly bear is one of the species he has tried to preserve, and he has won high marks in unexpected quarters for his efforts. "Ray has been good, very good about the grizzly," says Amos S. Eno, the wildlife coordinator for the National Audubon Society and a frequent critic of the present administration.

The most notable actions that indicate there may be a new era in grizzly management have involved law enforcement, no small problem. Since 1970, 42 grizzlies are thought to have been killed illegally in the Yellowstone ecosystem. (Since most such deeds are committed surreptitiously, there have almost certainly been more than are known.) Some grizzlies have been done in by irate ranchers and property owners who took the law into their own hands. Others have been killed by hunters who mistakenly—so they claimed—shot what they thought was a legal black bear. A number have simply been poached. A whole grizzly is one of the most valuable game trophies in the world—worth from $10,000 to $15,000. Also, in parts of Asia and in communities in this country made up of emigrants from those parts, there's a great demand for bear heads, hides, paws, claws and innards, which are thought to have powerful nutritional and medicinal properties. For example, smuggled into certain Oriental cities, a gallbladder of an American bear may sell for from $3,000 to $4,000 because it's prized as an aphrodisiac.

As a first and obvious step toward stabilizing the grizzly population, the interagency committee decided unanimously to make a coordinated effort to put a stop to criminal assaults on the bears. The gist of the decision was forcefully expressed by Arnett: "We're going to spend more money and manpower to prevent illegal killings. If we catch anyone who has molested a grizzly, you can bet your bippy we're going to come down on him like a ton of bricks."

To this end various states are making new law-enforcement arrangements. Idaho and Wyoming, for example, have altered their black-bear hunting seasons in the Yellowstone area so as to cut down on mistake killings. Despite a very tight Interior Department budget, the Fish and Wildlife Service last year got $200,000 extra for grizzly protection, an appropriation that doubled funds available for this work. FWS now has four mounted wildlife investigators riding the backcountry throughout the seasons when grizzlies and grizzly killers may meet, looking for evidence of illegal killings. (Because of the difficult wilderness logistics, it costs about $75,000 annually, including salary, to keep one of these agents in the field.) Also, last year some 60 other employees of various agencies spent part of their time, collectively some 7,000 hours, riding the bear range. Much of their work was educational—talking to ranchers, herders, loggers, hunters, outfitters, recreationists—explaining the grizzly problem, how to identify the animals, how to avoid trouble with them and the penalties for doing illegal things to bears. On the latter score, a direct reminder is provided by 12,000 posters now on display throughout grizzly country. These state that the National Audubon Society will pay up to $15,000 to anyone supplying information that leads to the conviction of an illegal grizzly killer. Since the poster campaign began last year, two payments have been made. The most recent, $4,500, went to an Idaho informant who tipped state game agents that one James H. Bibb of Priest River had poached a bear on or about May 1, 1982. Idaho agents obtained a search warrant and found that Bibb had in his possession various grizzly parts, including a full set of claws—much sought after for pendants and belt-buckle adornments. Also he had color photos—taken with a self-timing camera—showing himself at the scene of the kill, posing with his hunting bow alongside a dead 400-pound grizzly. Later it was determined that after making this portrait, Bibb cut off the animal's head and paws and packed out these trophies, leaving the rest of the carcass in the woods. Shortly thereafter, he displayed the photos to acquaintances in the Priest River area. (Poachers have great difficulty not bragging about their macho feats, a trait that often brings about their downfall.)

On being questioned, Bibb first claimed he had legally killed the bear in Canada; then he claimed he had found the animal dead on a road, stuck an arrow in the corpse and took the pictures so he could boast about them. Investigation demolished these alibis. Then Bibb said that he had indeed been bow-hunting for other game from a tree stand when a grizzly appeared and acted as if it was about to attack him. He killed it in self-defense. This story was also deemed fictitious because evidence disclosed that Bibb had been trying to lure in a grizzly with a rotten-meat bait. "It appears that he was never in any danger from the bear," said Dan Hawkley, an assistant U.S. attorney who in the fall of 1983 prosecuted the case in federal court in Coeur d'Alene, Idaho. U.S. magistrate Steve Ayers allowed Bibb to plead guilty to illegal possession of bear parts—a lesser charge than killing a grizzly. On Jan. 24, 1984, Ayers sentenced Bibb to a year in jail and three years of federal probation and fined him $10,000. The prison sentence and part of the fine were suspended. However, during his probationary period Bibb may not hunt and must volunteer 150 hours of conservation-related community service.

The penalties, the stiffest to be given a bear poacher, were widely publicized in the Northwest, and law agents believe the Bibb case will have a good, cautionary effect. There is statistical support for this optimism. In 1982, 14 grizzlies were known to have been illegally killed in the Yellowstone ecosystem. In 1983, the first for the come-down-on-them-like-a-ton-of-bricks policy, only six animals were lost in this way.

Since 1970 the body count of Yellowstone grizzlies has been officially tabulated as follows: 16 have died naturally or for reasons that haven't been determined; 42 have been illegally killed; and 135 have been lost through human actions that are considered legal and unavoidable. For instance, two animals were killed in bona fide self-defense situations, while six were traffic fatalities. The greatest number, 86, were victims of management, i.e., either intentionally or accidentally (most often by drug overdoses) done in by wildlife professionals while studying bears; destroying bears that posed clear threats to human life or property; or moving potentially threatening bears to more isolated areas.

These figures have made it clear to wildlife professionals that finding and managing habitats in which bears can live and not get into so much fatal trouble with us is the most critical preservation problem. This involves changing and restricting human behavior; bears can't be expected to modify theirs. "If we could treat the grizzlies with absolute benign neglect, completely isolate them from people, they would do fine and we could stop studying, managing and worrying about them," says Servheen, stating an accepted truism. If, for example, all of the Yellowstone ecosystem or northwestern Montana were completely closed to human use, the bears would almost certainly prosper. There are many other areas of the West into which bears could be transplanted and would thrive under similar arrangements. (For that matter, if northern California were evacuated and the weeds allowed to grow for a few years, it would make good grizzly country.) However, the expense and social commotion that would accompany any such action makes it fantastic. Therefore, grizzly bear planners and managers continue to ponder and argue habitat solutions which, if less promising for the animals, are more practical in human terms. Currently three alternatives are being seriously considered:

Modified Benign Neglect

The Interior services, Parks and Wildlife, favor regulatory changes that would cut down on the opportunities for bears and people to meet. To this end nearly 20% of Yellowstone Park was closed to visitors during parts of last year. Other temporary and permanent bear closures have been made in Glacier and in other parts of Montana. The creation of Situation I administrative districts where grizzlies more or less have the right-of-way is another step in the same direction. Whether this policy can—or should—be extended sufficiently to stabilize the grizzly population is becoming more of a debatable matter. Last December the Wyoming Outfitters Association, for example, passed a resolution opposing "the closure of large areas of Yellowstone Park and national forests to human use." Other commercial and recreational users have expressed similar opinions.

Habitat Improvement

Neither the Yellowstone nor Glacier area is particularly rich in bear resources, having become more impoverished because of developments made by humans. The animals hang on there not out of choice, so to speak, but because nothing better is available. Craig Rupp believes that an effort to enrich current bear habitats and create new ones would benefit the species and help to solve some of its human relations problems. Until his retirement in 1984, Rupp was the director of the Forest Service's Rocky Mountain region and a prominent member of the grizzly committee. He's one of the most knowledgeable advocates of the improved-habitat approach. (Within the service, Rupp also had a reputation of being a grizzly progressive. Many of his colleagues, at least until very recently, have tended to regard the bear and the hullabaloo about it as a colossal pain, something that interfered with their "real" work of leasing out grass, timber, mineral and energy resources.) Rupp says, "If the biologists, as they haven't done yet, would ascertain what is optimum bear habitat, give us a model of what it contains, we could manage forest areas to produce it—more huckleberry patches, meadows, downed logs for ants and grubs, the most desirable tree species, whatever. Having optimum habitat would presumably make the population more vigorous and would tend to concentrate the animals in these areas, keep them out of other ones. Some closures might be necessary, but I don't think so many restrictions on other uses are as necessary as some people claim. We could use logging, perhaps even grazing operations to help create and maintain the kind of habitat we wanted. The thoughtful use of such renewable resources can enhance rather than degrade an ecosystem."

Biologists agree that habitat improvement is a desirable long-term goal but don't feel it promises, in the short run, to alleviate the present crisis relating to the Yellowstone bears. It's also pointed out that all intensive-management practices increase rather than decrease the possibility of bear-human encounters. Though it's dear to the heart of foresters, multiple use isn't a theory for which grizzlies have ever shown much respect.

Supplemental Feeding

Essentially this involves a return to the pre-1970 situation in which many bears in the ecosystem hung around and fed at garbage dumps in Yellowstone Park. Proponents don't advocate reestablishing the pits as they were but rather setting up a series of feeding stations in isolated areas that would be supplied with good bear chow, possibly including quality garbage from the park. Scientist backers of this idea suggest that the Yellowstone animals may at least temporarily need such nutritional assistance and it might improve their reproductive rate, which declined after the dumps were closed.

Users of adjacent national forest lands—ranchers, outfitters and hunters—tend to favor supplemental feeding because it's assumed it will draw animals back into the park. If this happens it's hoped there will be fewer bear-inspired restrictions on people using other areas of the ecosystem. This line of reasoning has support at the highest levels in the Forest Service. In 1983 John Crowell, an Assistant Secretary of Agriculture, wrote a memo on the subject to Max Peterson, the chief of the Forest Service and Crowell's subordinate. After taking a passing swipe at the Audubon Society for its "very narrow and single-minded interest in grizzly bear preservation," Crowell said that the Service, of course, had to be concerned with broader, multiple-use matters. He said it seemed to him that most of the trouble, or at least the Service's trouble with the grizzly, came after the parks closed their garbage dumps. "Therefore it seems to me a top priority for the grizzly bear interagency team is to undertake reestablishment of such a core population [in the park] through artificial feeding, although not necessarily at dumps."

The Parks Service is adamantly opposed to supplemental feeding because, as a practical matter, it, like the other agencies, believes that if this policy is followed for any period, Yellowstone will once again have most of the bears and most of the bear management headaches in the ecosystem.

The federal Fish and Wildlife Service and centrist conservation groups, including the Audubon Society, feel that supplemental feeding should be used only in emergency famine situations and as a tactical management tool for moving—by baiting—bears within the ecosystem. As a regular practice, supplemental feeding is judged to be too expensive, usually unnecessary and a device which can be used by its advocates as an excuse for not making more important decisions about securing and improving natural habitat in which the bears can fend for themselves. Servheen says, "A lot of grizzly behavior is acquired, cubs learning from their mothers. If we habituate them to artificial food sources, they're going to be less inclined to forage on a mountain, perhaps forget how to. Their habits will change rapidly. We'll have something that looks like a grizzly but behaviorally will be something different from the wild animal we set out trying to preserve. If we have to feed them to save them, I, just as a personal, moral judgment, would rather see them disappear, go out on their own with dignity."

We're now probably spending more than $2 million a year to keep the token grizzlies we want in the West, although the total can only be estimated because it's made up of items that appear under different headings in many budgets. The off-budget costs are all but unaccountable in the conventional way but are also high: the price of closing 20% of Yellowstone Park; barring or restricting legitimate human activities in other areas; and the occasional property destruction and death that go along with keeping these animals.

We are at this time paying more to preserve grizzlies than we are for any other endangered or threatened species, but if we're earnest about getting the job done we'll no doubt pay more in the future. Clearly the grizzly can only survive in our midst as a managed species, something we have to keep accommodating, working with and arguing about for as long as it's here.

The grizzly controversy is generally cast in technical terms as a dispute about the best way to save the animals. However, the essential issue is cost: Bureau A and User Group B claiming that under Plan C they'll come in for an unfair proportion of the overt and covert costs; that we should think about Plan D, for which they'll pay less and somebody else more. But estimates of the true costs of having grizzlies have seldom been interjected into the public discussion, managerial types being understandably reluctant to be messengers bringing unsettling bills. Also, questions about price inevitably raise larger ones about values, and this has been a matter on which the bear establishment and the larger environmental community have been singularly mum. In the voluminous testimony about how to save grizzlies, next to nothing has been said about why have them.

After a routinely hectic day at the office, wrestling with bear papers and problems, Servheen talked one evening about the worth of the grizzly. The bear-is-ecologically-insignificant argument annoys him, and he makes an interesting rebuttal. He says the fear of bears cuts down on the number of people traipsing around in the backcountry of certain national parks and forests. Thus, the animals help prevent overuse and preserve the wilderness. This leads to a more intricate concept.

Servheen is a recreational user of grizzly country, a hiker, cross-country skier and hunter, though not of bears. He says that when he has had it up to the eyeballs with his job he likes to go away for a few days and pack into an isolated district in the Mission Range of Montana. There he's unlikely to meet any people but is always very aware that he might come across a grizzly. (Servheen carries a gun on these excursions.) "I'll hear an odd sound and prickles start on the back of my neck," he says. "It's fear, but the consequences are interesting. While I'm in a place like that my senses seem extraordinarily sharp. It's more than being alert to danger. It's a kind of super awareness about myself and everything around me. Those are the times when I feel most alive. If there were no bears I wouldn't have that experience, which I think is very valuable."

I obviously am much less intimately acquainted with grizzlies than is Servheen. However, once, to the northeast of Great Bear Lake in Canada, I slid down a pinga (conical frost boil) and only by dumb, but very good, luck caught myself on a scrub birch or I would have fallen on top of a female grizzly and her cub, which were 15 feet or so below in a rose thicket. Until that moment I had no idea that any bears were in the vicinity. Fortunately, the sow may have had little idea of what I was doing there or even of what I was, since this is virtually uninhabited, very rarely used country. Without even looking up, the female grunted a few times and with her cub padded off into the scrub. While they were leaving I was absolutely motionless, not as a reasoned tactic, but because for the only time in my life I fully understood the meaning of the cliché "petrified by fear." After they were gone, I think I experienced the sensation Servheen was talking about, one of marvelously heightened consciousness. Those few seconds remain the most memorable, the most powerful experience of a generally stimulating summer in the Arctic wilderness. In fact, except for a few people, some occasional lines of writing, maybe one or two athletic moments, nothing else has given me a high comparable to the one that the two bears provided.

That a grizzly can give such a direct, sensual charge doesn't in itself seem sufficient justification for making a complicated public effort to keep a few of the animals in our western states. There are simply too few who would place much value on being so gifted by grizzlies. Nevertheless, I'm solidly pro-bear and think they're generally worthwhile, despite the trouble and expense of keeping them.

First, in trying to save such an animal for and from ourselves we express critical, definitive elements of our humanity; we demonstrate that we can act on abstract principle for intellectual, esthetic and compassionate reasons. In such situations we also admit that because of these attributes we have unique responsibilities not to bust up marvels that we cannot recreate. All of our natural preservation efforts confer the same benefits, but the grizzly seems special because he's such a hard case. Keeping this beast alive will give us a real workout. We need such exercise. Insufficiently used, ethical reflexes go soft and flabby like muscles and then can't be depended upon in real crises.

Second, wilderness areas are gaudy demonstration plots of the ancient, immensely complex system that has produced and supports life in this world. We probably will never know exactly how the system works. But it's important—not just for good mental health but pragmatically—that we understand that we operate with and within the system. Hubristic conceits about resigning from nature or dominating it are dangerous delusions. When we try to act on them we are always slapped down, as a fish would be if it tried to hike the Appalachian Trail. To be reminded of the mystery of the origins of life, the brightest of us can get the point by contemplating a cockroach or gas station, but most of us need an exhibition like the wilderness to drive home the message. And the grizzly? He has such prowess, history and reputation that he isn't only of the wilderness but also makes wilderness on the ground and in our heads. We can sit securely at a scenic highway overlook and, if we know that there are great bears in the expanse of mountains and forests below, we understand immediately that we're in the presence of by-God wilderness. Valuable reflections about our places and possibilities often follow.

In 1691, Henry Kelsey evidently had some discussions with his savage companions about the value of the great bear. He reflected on them in the last lines of that first grizzly poem.

> His hide they would not me it preserve
> But said it was a god & they should starve.

We now express ecological observations much differently—and by and large far more prosaically—but the same principles still apply. This remains a wild place where we make meat for as long as we can, but eventually become meat. Trees or muggers can fall on us, we can succumb to microbes or motor vehicles, incinerate ourselves for political reasons. But the great bear is the only thing in these parts that now and then, simply as an expression of its nature, can step out and make us his meat. We can take the bear out of the woods permanently, but this won't change the reality he so powerfully represents—only make it more difficult to understand and remember. The grizzly is a valuable *memento mori*.

Nature Loving

Fifteen or twenty years ago, and for a long time previously, it was not uncommon for people to call themselves without shame and be called without derision nature lovers. These days you hear the expression less frequently. Now we have terrestrial ecologists, outdoor recreationists, and especially we have environmentalists.

One reason for the new terminology is that the contemporary environmental movement is concerned with many matters—invisible pollutants and poisons, complicated laws and regulations, social and economic trends which are too abstract and disembodied to serve as conventional objects of affection. Nature lover is simply not a suitable designation for a lawyer who spends his days trying to promote geothermal power and discourage the use of fossil fuels, even though one of his objectives may be to create better waterfowl habitat.

Also, the change in labeling has occurred because early on in the environmental wars of the past decade "nature lover" became something of a dirty term. Opponents took to calling those who pressed for environmental reforms and progress "a bunch of nature lovers." The implication was that these folks were silly sentimentalists, very likely secret Swedenborgians, who kept too many cats in their bedrooms and were inclined to wear runover tennis sneakers to congressional hearings. A good many self-proclaimed nature lovers did not help matters any by insisting that theirs was a mystical passion which required exceptional sophistication to appreciate and a transcendental poet to describe. In any event, nature lover, like knee-jerk liberal or arch conservative, became a label which serious-minded people wanted to avoid. Today .90 caliber professional environmentalists tend to buck and kick if anyone tries to brand them as nature lovers, and they themselves use the word with considerable sarcasm. "We've moved a long way from the bird, beaver, and babbling brook crowd," will say the geothermal lawyer. "We are more concerned with human behavior and survival than we are with the whooping crane. My God, just loving nature is not going to solve the environmental problems we face."

As a nature lover for as long as I can remember (and variously a terrestrial ecologist, outdoor recreationist, and environmentalist when politic), it seems to me that the debasement of this phrase, the disuse into which it has fallen, is unnecessary and unwise. In a broad, almost tautological way, nature lovers are those who take pleasure from going off into surroundings which are relatively unaffected by men and there variously sensing, experiencing, and contemplating "natural" phenomena. (Common usage but certainly neither logic nor science has it that man is somehow unnatural, but that is another paradox, which has been addressed previously.) Obviously a lot of us do this sort of thing and can better be called nature lovers than something fancier.

Besides being unseemly to call a spade a manual excavating tool, trying to suppress the fact that many of us are simply nature lovers obscures and distorts an important truth about our present environmental concerns. That is, that a good many are motivated by affection for certain wildish lands, an attractive body of water, an interesting plant or animal species, rather than by fear of the environmental apocalypse. Underestimating and deriding this affection has been a chronic error of those who oppose "environmentalism." Environmentalists may be making much the same sort of error if they try to disguise the affection with harder, more elaborate labels or convert it into fear. Art lovers do not create art but they do support artists, museums, and galleries. Analogously, nature loving may not solve intricate environmental problems, but nature lovers, affectionate rather than fearful people, are generally the ones who pay the bills of and cast the votes for the environmental movement.

All of which recently came to the surface as it does now and then for me. For reasons that do not signify here, for six months or so I have been traveling around the country talking to environmental professionals. I learned a lot, but I became stuffed to the gills with abstraction, jargon, and computer printouts. (This is not to suggest in a backhanded way that environmentalists are exceptionally tedious. The result probably would have been the same if I had consorted excessively with theologians, chemists, or, God forbid, magazine journalists.) Consequently I felt the need for a little sensual experience as a spring purge. As March waned and the shadblow began to bloom, the feeling intensified.

Fortunately, two longtime companions were in the same mood. Ky, my son, had been working all winter in the woods and was tired of fighting balky chainsaws and trying to plant pine seedlings in rocky ground. Sam, a close mutual friend, is a young, creative farmer and a very new father. He had been trying desperately to raise enough money to buy the small farm he rents so he could turn it into an experimental, high-yield orchard of exceptionally dwarfed fruit trees. During the previous month he had helped to deliver his first son, spent a lot of time with bankers and lawyers, and entertained his mother-in-law as a house guest.

The three of us met one afternoon over a rock patch of a streamside garden which we were trying to plow. Between cursing the rocks and beating on the senile tractor, we decided we needed to get out and away and do something quickly. For the three of us, doing something has always meant heading out into the bushiest place available. In the past, as energy and resources permitted, we roamed the Arizona mountains for a year watching coatimundis, dragged canoes for a summer through the muskegs and mosquitoes of the central Arctic, and did some fairly fancy spelunking, rapids running, and climbing. However, at this particular time we were not looking for anything ambitious or complicated. We decided the best thing would be to strap two canoes on the van and drive into northwestern Pennsylvania until we found a likely river we could float easily while looking around the countryside.

This is exactly how things worked out for us. We had no enormous adventures, encountered nothing especially rare or spectacular, went no place where other people have not gone or where they could not go if they were of a mind. Nevertheless we had a fine, purgative few days. Also, in retrospect the outing provided some typical and specific examples of what is involved in nature loving, and they may be worth reporting as contribution to the campaign to clean up and restore the good name of this old and honorable activity.

One of the most obvious and significant things about nature loving is that a lot of people do it in a lot of different ways. We began to collect evidence supporting this premise even before we left for our river.

Trout season opened in Pennsylvania the morning we left. The stream that runs past the abominable rock garden patch is well stocked and each April draws a considerable number of fishermen, many from outside the immediate area. One who has been coming for almost a decade is an elderly gentleman, beautifully outfitted with old, worn gear which has been lovingly preserved. Sometimes on bad winter nights I have thought of him, fantasizing that he is sitting before a fireplace somewhere, a light drink by his side, happily fussing with his fishing paraphernalia.

We talk a bit each spring but never have formally introduced ourselves. Because of his dignity, formal and courtly manners, I have guessed he might be a county seat attorney, or perhaps a judge. For the past two springs he has come with his grandchildren—a boy and girl both under twelve—and he spends more time instructing them, untangling their lines from sycamore limbs, than he does fishing himself.

The judge, if that is what he is, and the children arrived as we were lashing down the canoes. Though long ago he was made perpetually welcome to our section of stream and bank, he formally requested permission, as is his custom. (This alone makes him memorable among trout fishermen, who in manners and deportment seem each season to be getting more like deer hunters.)

"Then, sir," he concluded our annual conversation, "with your kind permission I will take the youngsters down to the hole below the bridge."

"My daughter got a couple there yesterday, but there should be more."

"Indeed there should be. It is such a fine place for beginners— open enough so they avoid difficulties but sufficiently productive to hold their attention. As you can observe, my principal interest now is in helping them. An old party like myself has taken enough trout for a lifetime, and a few more or less is unimportant. I carry the rod as an excuse to get out with the children, walk along the stream, enjoy the surroundings, and think back of the happy hours I have had in such places during the past sixty years. Again sir, my gratitude for your hospitality."

"You are more than welcome, and we'll see you next year."

"It will be my pleasure, but at my age long-range commitments cannot be guaranteed."

The old trout fisherman was the first of a number of similarly motivated if variously engaged people we saw that morning while driving northwestward across the state toward the river of our desire. With the larger figures being only estimates—since we did not start thinking about what we saw until later—the tally is as follows: 300 trout fishermen; 25 canoeists afloat in various rivers, and about twice that many cars carrying canoes; 30 bicyclists; 25 backpackers; a dozen horseback riders, and perhaps 20 cars trailering horses to trail rides or shows; eight morel mushroom hunters; five rock climbers; four men scouting in the mountains in preparation for turkey season, which would open the next week; two boys and a man using binoculars to watch ducks in the Juniata River; a watercolorist and her companion, set up along the berm overlooking a fine display of dogwood and redbud; one Indian arrowhead hunter. Also, we passed parking areas adjacent to two state forests and a park. They contained perhaps a hundred automobiles whose owners were off in the woods doing who knows what. Among the motorists we passed there was no way of knowing how many were driving around simply to see the scenery and new foliage, getting ready to picnic or sit in the sun. Considering the season and the warm day, a good many must have been out for this reason. All of these were brief, accidental observations made casually while pushing along at 50 mph on main highways. Had we scouted the backroads or gotten out and walked a bit, both the numbers and categories of nature lovers certainly would have been multiplied.

About ten years ago I spent an afternoon with Starker Leopold in his office at the University of California. A distinguished zoologist who has had an important influence on public wildlife and conservation policies, Leopold is also a member of perhaps the most influential environmental family in the country. His two brothers and sister Estella are respected authorities in various natural history fields. His father,

Aldo Leopold, was the author of *A Sand County Almanac*, often cited as the seminal book of the modern pop ecology period. During our visit I asked Leopold if he thought the number of nature lovers justified, in a social sense, the amount of public land, money, and effort devoted to various conservation projects. He answered, according to an entry in a yellowing notebook, "All available records, park and forest-use logs, recreational surveys, indicate that the overwhelming majority takes some sort of interest in, pleasure from, natural history experiences. So far as I can see, only a very tiny minority do not. You travel around, talk about these things—how often do you hear people complain that we have too much wildlife, too much natural scenery, too many woodlands, wetlands, undeveloped rivers, lakes, beaches? For that matter, when was the last time you met a nature hater?"

Leopold's statement has always seemed to me to be one of those simple and obvious (after the fact) comments which become more profound the more they are considered. I thought about what he had said again that morning while zipping past the miscellany of Pennsylvanians who were out enjoying their surroundings. Nature loving may not be universal, but far from being an elitist avocation it must be close to being our most popular form of recreation, the commonest sort of esthetic activity and a very general source of satisfaction.

As we drove northwest we decided that the river which best suited our immediate needs was the West Branch of the Susquehanna, and that the place to put in was Karthaus in Clearfield County. From there we would drift twenty miles or so downstream to Keating at the confluence of Sinnemahoning Creek.

Karthaus is a community of two or three hundred wedged between the river, railroad tracks, and coal tipples colored gray with ground-in coal dust. Twelve years ago, the last time I was there, it seemed on the verge of abandonment, as did many villages in the region. Nearby mines had closed or severely cut back production. Remaining residents had to drive fifty or a hundred miles a day to work when they could get work. Many of the houses were abandoned, and the principal building, a hotel-tavern, was on the verge of extinction.

Currently Karthaus, though not a beauty spot, is much livelier and more prosperous than it was when I last saw it. It has been revitalized by distant happenings over which local residents have had no more control than they do over the phases of the moon. The Arab oil embargo, the subsequent energy shortage and panic, have made it profitable to dig soft coal again. In consequence there are new cars in the streets of Karthaus, spruced-up houses, and a dozen or so afternoon patrons in the tavern where the Iron City flows and the pretzels crunch. Even so, a little inquiry located a resident who for ten dollars and the novelty of it was willing to follow Sam down to Keating, where he left the van, then drive him back to Karthaus, where we would begin to float down the river.

After they returned, Sam said of the driver, "He's a strip-miner, and he says things have never been better around here. They're digging coal, and they have to restore the land after they strip it. He says it costs his boss $1,700 an acre to fill in and replant after they've finished with a pit, but this puts money in the pockets of the working man. The best thing is that they're selling their coal directly to the Japanese for fifty dollars a ton, which is twice what they get here. They have to because the coal is too high in sulfur to pass our environmental regulations. He might not want one of them to marry his sister, but otherwise he is all for the environmentalists and the Japanese."

Between Karthaus and Keating the West Branch is fifty to a hundred yards wide and flows through a fairly tight gorge, being contained by ridges that rise 500 feet or so above the water. Except for occasional deep pools the depth is seldom more than three feet. However, enough water comes in from mountain springs and creeks that this level is maintained even in a dry season, as this spring has been. There is a good pitch to the river, which moves along at the rate of a mile or two an hour. Floating such water in a canoe may be the most restful and pleasant of all modes of transportation. It is noiseless, nearly effortless, as safe as progress can be. You move fast enough so as not to become fidgety, but slow enough to get a good look at everything along the way.

The West Branch is a remarkably clear stream, clear enough to be used in a beer or cigarette commercial where the intent is to equate the purity of the crystalline stream with the product. However, if the West

Branch were to be so used, the background would be as false as the message. The water of this river is clear by reason of corruption, not purity.

It does not take long or any extraordinary powers of observation to discover that something is very wrong in the West Branch. Though the visibility through the water is almost perfect, there is almost nothing to see below except rocks and debris. There are no fish, no crawdads, amphibians, rarely an aquatic plant. If you scoop up a handful of river and taste it, it tastes bitter and sour, as it should since it is in fact a mild sulfuric acid solution.

Along the West Branch and its tributaries above Karthaus are old mine shafts and pits which have filled with water. The water has dissolved the sulfur in the exposed veins of coal and the coal waste; welling up out of the pit holes, this toxic solution pours into the West Branch. The acid wastes have stained the riverbanks, painted the rocks and mud a bright orangy-red. In this wasteland one of the few plants that thrive is the acid-loving cranberry. The situation of the cranberries suggests a metaphor illustrating one of the fundamental but often intentionally overlooked facts of ecology. Ecology is the study of the relationships between living things and between living things and the inanimate environment. These relationships are *always* changing and *never* static. Everything is in the process of becoming something else, which is necessary for the maintenance of our planetary life system. However, in the short term, changes work to the advantage of some of us, to the disadvantage of others. The contamination of the West Branch with acid mine water has been "bad" for trout, crayfish, and some men (say, trout fishermen, who have no economic interest in coal). On the other hand, the acid cleared the banks of competitors and opened up promising new territories for cranberries. Ecologically the effects of change can only be assessed in terms of our own short-range self-interest—absolute and universal value judgments about changes are *always* fantastic.

The wasted West Branch does not support much life now, but considerably more than it did a dozen years ago. Then it was virtually sterile, but on this trip we found some new mats of aquatic plants in places where freshwater streams were emptying into and diluting the river. In one of the plant beds there were a few periwinkles. A fisherman we met on a tributary creek said he had seen a few minnows in the river. The acid-stained belt along the banks has narrowed. Sedges, ferns, and other plants are pushing into and displacing the cranberries (for whom the Golden Age of Acid may have passed). On this spring day there were rafts of ducks, mostly woodies, mallards, and goldeneyes, floating on the river, and at the edge of the water raccoon tracks were plentiful. Overhead there was almost always a circling osprey. We did not see any of these creatures take food from the West Branch. The birds probably were migrating elsewhere, and the coons may just have been messing around in the mud as they will. Nevertheless, there was a feeling that a lot of life was crowding in closer to the river, testing it, sensing that it might be becoming habitable.

Just as large economic and diplomatic events have caused the rejuvenation of Karthaus, so the tentative revitalization of the river is the result of complicated happenings which have occurred far from this body of water. That the West Branch supports a somewhat more diverse biological community than it did a decade ago reflects—in an ascending order of specificity—the facts that: we have committed $18 billion of public funds to cleaning up our waters; substantial federal and state funds have been spent to determine how to abate pollution from coal mines; for the past five years the State of Pennsylvania and the Barnes and Tucker Company have been locked in a bitter legal wrangle over who should pay to stop sulfur-polluted water in one of the company's abandoned mines from pouring into the headwaters of the West Branch. While this dispute—there are similar ones—continued, the courts enjoined the two parties to split the cost, about $35,000 a month, of detoxifying the water and pumping enough of it out of the mine that it would not overflow into the river. A few days previous to our float trip the Commonwealth Court decided that Barnes and Tucker must pay all past and future costs connnected with keeping the acid water from their mine out of the West Branch.

The environmental establishment has rapidly become very large and increasingly institutionalized and is now almost as big and cumbersome as the military bureaucracy. In consequence it is now possible and fashionable to be critical of its red tape, waste, contradictions, and timidities. Yet this does not alter or obscure the salient fact that within less than a decade our environmental concerns, policies, and to be fair, bureaucracies have measurably improved the quality of life. For example, 97 percent of all major industries that discharge waste into our waters now operate under a permit and inspection system which eliminates or drastically curtails their water-polluting activities. Few such restraints existed ten years ago. There is 25 percent less sulfur dioxide and 14 percent less dust, smoke, and soot in our air than there was five years ago, and appreciably less DDT in our body tissues.

Nobody can seriously doubt that we have horrendous environmental problems which, if ignored, will probably make life increasingly ugly, disagreeable, and uncomfortable, if not impossible. However, environmental successes already achieved indicate that catastrophe is not inevitable. Given sufficient resources and self-discipline, many of our problems can be minimized if not solved. The West Branch is a microcosm of this situation. It is a sick river, largely unfit for man or beast. In this respect it is a frightening place, demonstrating how thoroughly we can foul our own and other nests. But the fact that some water weeds and periwinkles are making a go of it where they could not ten years ago suggests that even such a sick river can in part be restored and that we have shown some interest in doing so. In that respect the West Branch is a hopeful place, suggesting that we have some control over our fate.

One of the slanderous inferences about nature watching, appreciation, loving, or whatever, is that it is a heavy, solemn activity, vaguely uplifting and good for you but also stuffy and tedious, comparable to sitting through a long sermon or reading the collected works of Henry James. In fact, floating a river or doing something similar is in many ways more like being in a busy, lively bar where all sorts of topics are being bounced off one another in instructive and entertaining ricochets. Without much effort on your part you are constantly encountering phenomena which stimulate speculation and, if you are with somebody, conversation.

For example, because of the general situation along the West Branch and in Karthaus we chewed away for the first few miles on such matters as the price of coal in Japan, the effect of the Alaska pipeline on breeding peregrine falcons, the internal politics of the Environmental Protection Agency. By and by we pulled up along the bank and climbed up on a low, heavily wooded knoll with the intention of drinking a round of beer. In this thicket we found a single, sickly looking locust tree whose presence and condition touched off a new round of conversation which was tangentially related to the previous Karthaus-coal-acid-environmental discussion.

Coincidentally, locust trees were a matter of some immediate interest to Ky and Sam. Locust is the most rot-resistant of all our native eastern woods and therefore is sought after for fence poles. A seven-foot-high, four-inch-in-diameter locust post presently sells for $1.50 to $2. In a small way both of them were in the business of cutting and peddling locust poles and had plans to expand their activities. Sam, in fact, estimated that the farm he was trying to buy had two or three thousand dollars' worth of locust standing around on the hoof, and this resource figured heavily in his elaborate financial schemes aimed at buying the place. Additionally, as we had driven toward the river we had seen several places where old strip-mining pits had been filled and planted with locust seedlings. This seemed to us to be a good innovation, better than the practice of reforesting with rows of pines. The locust is the tree which naturally would appear first on denuded land, and in time locusts should produce a forest more suitable to the Appalachian hills than one started as an artificial pine plantation. In any event, locust trees were on our minds, and when we found this lone one growing above the river we sat down, drank our beer, and considered the properties of the tree for a half-hour or so.

Typically the locust is a pioneer species, moving into and thriving on barren, open ground. However, the knoll on which we were sitting was heavily wooded, shaded with the kind of mixed deciduous forest that dominates much of the West Branch Valley. Beech, oak, maple, and a few white pines and hemlocks were the major species, while dense thickets of rhododendron filled in below. The single locust tree was out of place in this community and, from the poor look of it, was not going to survive much longer. Looking about, we saw evidence that this bank of the river had been timbered—probably fifty years previously. The logical assumption was that after it had been cleared, locust had moved in and done well. The tree before us was presumably the sole survivor from the Locust Period, now hanging on accidentally like a dinosaur in the Age of Mammals.

A locust can grow in poor, thin soil, can get along without much water or mulch, can stand, in fact thrives in, the blazing sun. Tough and thorny, the locust is well able to defend itself against browsing animals who also like clearings and will trample down and gobble up less well armed seedlings and saplings. Locusts will appear in old fields and pastures which have been cleared for and eroded, leached, and nutritionally exhausted by agriculture. Locust moves in after fires, after logging, and now after strip-mining.

As soon as locusts appear in poor places they set about enriching them. Their roots stabilize what topsoil remains, break up the hardpan and improve its capacity to retain water. Fallen locust leaves contribute mulch. Most important, the locust is a legume. Like other members of this family, it has the useful capacity of taking nitrogen from the air and transferring it to and thereby fertilizing the soil. After some years of this sort of thing the land is made habitable for larger, grander, but—at least in their infancy—more delicate trees: the beech, maples, oaks, evergreens which were growing on the West Branch knoll where we sat drinking beer. Seedlings of these species mature, and in time they repay the pioneering locusts by cutting them off from the sun, crowding them out, eliminating them from the ground.

All of which is very tough cheese if only the self-interest of the locust is considered; but that, the interest of one species or community, does not seem to be the point of our intricate life-support system. The system functions, or always has, so that while species are constantly struggling to achieve permanent domination (that is, create a static, changeless situation which they are best able to exploit and monopolize), none is successful in this object. They fail because of the competition of other living things, but also because the closer they come to dominance the more vulnerable they are to their own excesses. Both the rise and demise of locusts perpetuate this system, and in the long run perpetuate the locusts themselves.

In the proper mood one can find and be enchanted by complexity almost anywhere. The process which delivers a daily newspaper to the door and the system which enables us to eat breakfast in New York, cold, soggy pastry over Kansas, lunch in Los Angeles are awesome examples of intricacy. However, no place are wonders of sheer complexity so well and obviously exhibited as they are in the natural world. You can pick up any single strand in what has been called the Web of Life—a locust tree, a periwinkle, the sulfur content of western Pennsylvania coal—and if you care to follow it far enough you will find it is connected with every other strand. This kind of following and finding is a dependable, always available satisfaction of nature loving.

Web of lifing is by no means the only reward of or the only thing to do while cavorting about the fields and streams. Simply admiring a thing for what it is, no matter how it got there or where it is going, is equally enjoyable. In this category, red-winged blackbirds have always been for me one of those very-good-in-themselves phenomena. I grew up on the edge of an extensive Michigan marsh in which I spent a lot of my free and formative years. I did not usually go into the swamps solely to admire blackbirds. (Except at the very end of this period, I was generally more interested in mink, muskrats, and snapping turtles and in building secret huts.) However, they were always around, providing a pleasant background. Their looks—like smart, well-turned-out guardsmen—have always pleased me. Their behavior pleases me; they are bold, self-confident, up-front creatures. A redwing will rise out of a marsh, chase away a much larger hawk or crow, but go back about his business without spending the whole day admiring himself or bragging about what a great hero he is. I like where they live—in wettish, tangled places—and especially I always have enjoyed how they sound; my favorite bird music is the *oka-lee, oka-lee* of the redwings' spring song.

    I have encountered redwings in hundreds of places. Perhaps the ones I liked best I met on a cold, blizzardy day in June while I was paddling in an ice-choked subarctic river between Great Slave and Great Bear lakes in the Northwest Territories. To my astonishment I came on three redwings hopping about on an ice floe, presumably picking up stranded aquatic insects and larvae. Before I passed they did a little *oka-lee*-ing. This did not settle my problems, but it was soothing. I figured if this country was suitable for redwings, it must have something to recommend it—and it turned out it did.

Because of the canyonlike terrain and contamination, the West Branch does not have much in the way of adjacent swamps. However, we did come on one largish tributary stream which had formed a mini-delta and created a few mushy, marshy acres at its mouth. In this place there were some redwings, and as we walked through the mud I watched, listened to, and loved them. By and by I did my redwing appreciation number on Ky and Sam. Not having heard it for several years, it was high time they listened to my account of blackbirds on the ice floe again. Then I told them my best redwing story, which surprisingly they had not heard before.

"After we were married, we spent the year in southern Mexico. We ran out of money and came back to Michigan in April. We opened up the cottage at the lake and more or less camped there. We had a temporary job running a kids' camp in Virginia, but that did not start until June, so we had six free weeks. Everyone thought I should spend the time filling out job applications. My mother was very hot on the subject. She sat me down one day and said I absolutely had to find real work—I was twenty-five years old and had made only $150 writing and should be ashamed of myself. What we actually did was go off every day into the marshes behind the cottage and watch redwings. There were about 25 pairs in the section we patrolled. We kept a fairly detailed record of nesting and hatching dates, the development of the nestlings, feeding behavior, and mortality. I still have those notes someplace. We never published them, and probably there was no reason to do so. Redwings are easy birds to observe, and chances are anything we found out a lot of other people had found out years before. But we had a great spring, and in fact got so interested in what was happening in the marsh that we were a few days late getting to Virginia for the camp job."

"So," said Sam, " you owe your subsequent fame and fortune to sloshing around watching redwings."

"No, but that helped to make it obvious that fame and fortune were not going to be my movie, and that it would be inappropriate to pretend otherwise."

In the late afternoon we glided up to a high, flat, wooded bench on the south bank of the river. We hauled up there and looked around to see what it offered as a camping place. It offered a lot. There were substantial piles of beech leaves, very suitable for do-it-yourself mattresses. Some previous paddlers or hunters had dragged a lot of rocks up from the river and made a good fireplace. In the flotsam and jetsam at the water's edge we found a somewhat bent but sound saucepan we were glad to have, since it permitted us to boil coffee water in something other than a greasy skillet. Downstream a few yards we came upon an extensive patch of ostrich ferns whose fiddleheads were just emerging and were in a prime state of tenderness and thus edibility.

One way to fix fiddleheads is as follows: Pick a hatful of them. (Don't worry too much about the ferns; new shoots will shortly replace the harvested ones.) Scrape off as much fuzz as you can. Cover bottom of skillet with water. Bring water to a boil. Put fern shoots into water for a minute or two. Remove fiddleheads, dry them on a T-shirt, and set them aside on a clean canoe paddle. Fry four strips of bacon in skillet and hardboil three eggs in found saucepan. Peel eggs and mush them up with bacon and fiddleheads. Add as much sweet onion as you thought to bring along, and as much cheese as you can spare from the next day's lunch. Pour bacon grease over all, and you have something very tasty.

This improvised fiddlehead salad or stew was so successful that it seems worth making the recipe public, even though doing so may be a federal offense and an officially antisocial act. On April 29, 1976, after months of deliberation, the Federal Trade Commission announced a ban on advertising the use of "wild" foods. Such advertising is now punishable by a $10,000 fine. The test case which brought this matter to a head involved Post Grape Nuts TV commercials featuring the late Euell Gibbons (incidentally, a longtime Susquehanna river rat) eating and loudly enjoying things like cattails and blackberries. These ads, and others like them, cannot be used anymore because, according to the FTC, "of their tendency to influence children to engage in dangerous behavior." The Gibbons commercials, decreed the regulatory body, "undercut a commonly recognized safety principle, namely, that children should not eat any plants found growing or in natural surroundings, except under adult supervision."

Governments have a great talent for making it hard to love them consistently. Besides talking and writing gobbledygook, they are forever getting their right and left hands mixed up. While one squad of G-men is doing something admirable and gutsy like making Barnes and Tucker stop dumping acid in the West Branch, another bunch starts bad-mouthing Euell Gibbons. Probably the situation is insoluble in any abstract, rational way. Government by law, almost by definition, precludes common sense, since only individuals have common sense. Yet individuals unrestrained by law have a long history of being dangerous nuisances. However, in this particular case things may work out better in practice than in theory. Given the notable inclination of American youth to rebel against arbitrary institutional restrictions and discipline, the FTC edict may have the effect of sending hordes of youngsters off to nibble on wild things more or less as a defiant antiauthority gesture. What the FTC may really have done is create a new generation of blackberry junkies, cattail pushers, and fiddlehead addicts.

There are a good many places in the Appalachians which are wilder and less settled now than they were fifty or a hundred years ago. The reason is that the land was difficult to begin with and was used so badly that it no longer will support many people in the style to which we have grown accustomed in the latter part of the twentieth century. The West Branch between Karthaus and Keating is one of those places. Prior to World War I, and for some time previously, there were at least three small communities in this stretch, five public houses, and by report six or seven hundred permanent residents. Timber cutting, mining, railroading, small farming, hunting, and moonshining were the principal elements of the local economy. Now there are only seven families living year-round in the valley, and six of them are concentrated in Cataract, which was once a village of several hundred. Its reduced remains stand by the river on a boulder-choked bend which in the old days was called Buttermilk Rapids.

There is a man now living in Cataract who was born there around the turn of the century. He lives in a house his grandfather built three years before the flood of '89. Where he now has a garden patch, his grandfather's hotel once stood. The man says he can remember when the hotel was packed every night and you could walk across the river, stepping from raft to raft. (Before the railroad, upriver people would knock together timber rafts for floating produce to downstream markets. After the rafts had been unloaded, they were sold to sawyers.) The old man says he remembers when 150 Italian stonecutters camped on the mountain above Cataract and worked a quarry there. Cataract still draws its water from a spring-fed reservoir, the walls of which were joined a half-century ago by the Italian artisans. He remembers when the mines opened and closed, when the railroad was built to haul out the coal, and how it declined.

A mile above Buttermilk Rapids we stopped to look at the remains of an even more ghostly community—Old Belfonte. The only surviving building was once the hotel, a gingerbready structure that has the look of a small-town railroad depot, which it may also have been. Now it is a camp used by a group of hunters in the fall. The bench upon which the village once stood is still partly cleared and stands as a meadow set with a few old apple trees that were in bloom. It was such a pretty spot, and we are such good Americans, that we got to talking about how nice it would be to buy the building and the meadow, possess them as a retreat.

"All we need to do is come back in the fall and look at the apples and find a super one," said Sam, our orchardist.

"How so?"

"That's how you get new commercial varieties, just running across them in old orchards. They're accidents. Somebody found both the Red and Golden Delicious that way, under some other trees. Stark Brothers is supposed to have paid something like $100,000 for the Golden Delicious. The original tree is still growing in a cage. At least indirectly all Golden Delicious came from it."

"What happens if the seeds from that tree germinate, grow into trees? What kind of apples do you get?"

"You can't tell, but they wouldn't be Golden Delicious, and they probably wouldn't be very good. Developing hybrids takes too long in apples. That's why they look for accidentals and then take grafts from them. Right now everybody is trying to find a replacement for the

York Imperial. You still get the best storage apples from a York, but now they're infected with something called York Spot, some kind of blight. You waste a lot of the apple peeling away the spots."

"So all we have to do is come up here next fall, find a good York, and live happily ever after?"

"With my luck what we'd probably find is that these old trees are where York Spot got started."

In early afternoon two red-tailed hawks rose out of the woods and above the canyon rim and against a thunderous sky performed a marvelous dance. The birds of prey are a great passion for the three of us. We will stop whatever we are doing to look at and love a hawk, falcon, or eagle.

In cities, in man-dominated scenery, there are admittedly marvelous shapes and patterns which in no way are formally inferior to anything you may see in the wild. However, it seems to me they are generally harder to observe and appreciate because there are so many of them and they are so disparate. On a street your senses are constantly bombarded by competing stimuli: moving traffic, blowing horns, colored signals, restaurant smells, the commotion of an accident, time and temperature spelled out in blinking electric lights, high-rise buildings all of different designs, recorded music, strange faces, a stray dog, blowing wastepaper, a fat man eating a triple-dip chocolate ice cream cone. Our receiving apparatus cannot cope with such diversity. Powers of observation are numbed, we find it difficult to focus on details and must settle for a generalized impression of a complex scene.

In less busy and occupied parts, say, floating an easy river, the tempo and pace are much slower and the scene is more homogeneous. The black clouds, the ridgetops jagged with tall pines, the sound of running water and rising wind were a complementary backdrop for the dancing hawks, did not compete with their act or distract our attention. We lay down our paddles, let the current take us where it would, and gave our undivided attention to the hawks until they went a-courting over the mountain.

Shortly thereafter the thunderheads against which the redtails had performed fulfilled their promise. A cold, stinging rain began to fall. The wind turned and gusted into our faces. Only a masochist or some other kind of damn fool would honestly claim to enjoy paddling in a chilling drizzle into a headwind more than floating down a river under a warm sun with a gentle following breeze. However, bad weather is not all bad. For one thing, it gets your attention in a way good weather does not. Rain beating down, collecting as bilge in the canoe, seeping under the hood of a rain jacket, is in an odd way more stimulating than fair weather. At least in moderation, shivering is more exhilarating than sunbathing. In bad weather there is a feeling of being deeply, if contentiously, engaged with the elements.

Also, bad weather is a good reminder that we live in a kind of yin-yang world in which we are often supported and transported by bonded opposites. If you have not been frightened, you cannot really savor security. You will never be found unless you have been lost. Where is up if there is no down? In the same way, unless you have been soaked and chilled, become a little hungry, tired, and cramped, it is hard to appreciate how good it is to see a van parked along Sinnemahoning Creek; how good it feels to crawl out from a canoe, put on dry clothes, eat fried chicken, mashed potatoes, green beans, cottage cheese, and German chocolate cake in a warm railroaders' restaurant in Renovo, Pennsylvania.

Everybody will do things a little differently, but in general this is the kind of thing you do when you go out nature loving. It is quite simple and straightforward. It has never seemed to me to involve activities that are too silly, sentimental, or subversive to be admitted publicly.